BEYOND T~~

LOST AT SEA, NEAR ..., MAGIC,
DREAMS, AND GHOSTS IN TH ...ORLD OF PACIFIC ATOLLS

BILL OTEY

outskirts
press

Beyond Time
Lost at Sea, Near Death, Salvation, Magic, Dreams, and Ghosts in the Water World of Pacific Atolls
All Rights Reserved.
Copyright © 2022 Bill Otey
v2.0

The opinions expressed in this manuscript are solely the opinions of the author and do not represent the opinions or thoughts of the publisher. The author has represented and warranted full ownership and/or legal right to publish all the materials in this book.

This book may not be reproduced, transmitted, or stored in whole or in part by any means, including graphic, electronic, or mechanical without the express written consent of the publisher except in the case of brief quotations embodied in critical articles and reviews.

Outskirts Press, Inc.
http://www.outskirtspress.com

ISBN: 978-1-9772-4230-3

Cover Photo © 2022 Bill Otey. All rights reserved - used with permission.

Outskirts Press and the "OP" logo are trademarks belonging to Outskirts Press, Inc.

PRINTED IN THE UNITED STATES OF AMERICA

To The Memory Of Carlos Viti

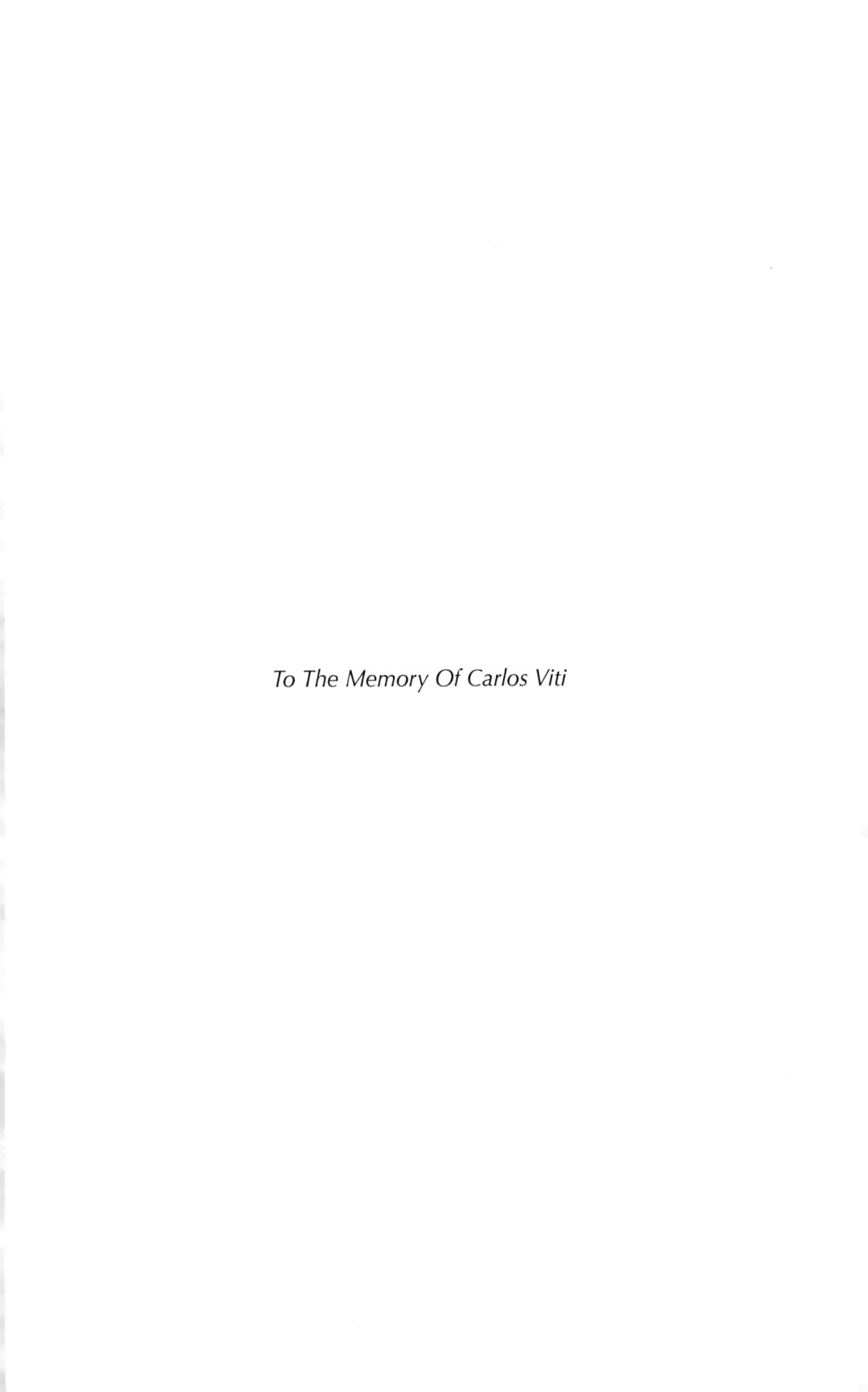

Acknowledgements

To Carol Ingram, former Peace Corps volunteer in Truk, who read my manuscript and got me straight with the English Language.

To Eric Metzgar, Phd, at Triton Films, who in many phone calls and email conversations, gave me valuable insights and got me straight on Micronesian Voyaging, Navigation, and Ethnography, and especially for helping me make a spirit connection to my work making this book.

To Dan Baker who knew Carlos well and gave me details of his life from 1973 to his death in 1996, and for permission to quote from his eulogy for Carlos.

To my children, Kathleen and Bryan, who will learn what Dad did in the War.

To my patient wife, Penny Suzette, who gave me the time and encouragement to finish this book.

To Ruby-Mata-Viti for kind permission to use photos taken by Carlos Viti.

To Karinjo Devore for kind permission to use the photograph of Yaleilei, by Nicholas Devore III which appeared in *National Geographic*, December 1974.

To Harvard University Press for license to use the photo of Hipour, from *East is a Big Bird: Navigation and Logic on Puluwat Atoll* by Thomas Gladwin.

To Monica Harman and the great team at Outskirts Press for their assistance and patience with a novice writer in putting this book together.

MAP OF REMOTE OCEANIA
The Voyaging Universe of the Carolinian Navigators

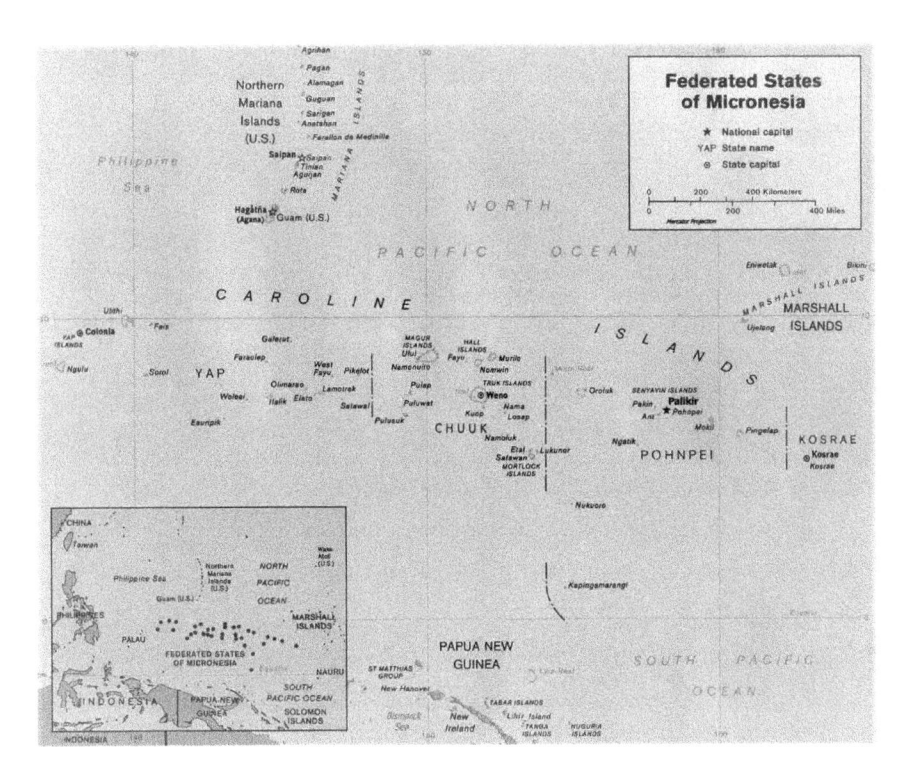

BEYOND TIME
Lost at sea, near death, salvation, magic, dreams and
ghosts in the Water World of Pacific Atolls

Table of Contents

Prologue i
The Maramar Hotel 1
Carlos 3
Pulusuk 10
The Way Of Wood 15
Perfect Evolution 17
Magic and Awareness 19
Stranger In A Strange Land 21
Ghosts 23
Father Fucky 24
Rivalry 26
Leaving Pulusuk 28
Water World 30
Tri Amora 32
Rock Stars 34
Provisioning 37
Leaving Truk 39
Sleighride 41
Puluwat 43
The Way of Turtle Dying 45
Satawal 46
Gift of Magic 47
West Fayu 49
The Shark 51

Over the Reef 57
Lamotrek 60
Dead Reckoning 63
Whale Watching 69
Near Death 70
A Dream 72
Salvation 74
The Island 77
Legacy of a Peace Corp Volunteer 78
Epilogue 79
References: Micronesian Ethnography and Voyaging 83
Photographs 85
End Notes 107

Prologue

This is a small story of events that happened in the Caroline Islands[1] of Micronesia a long time ago, Beyond Time; of a fateful voyage and the characters who will long live in my memory. The telling of this story began when I stumbled onto exchanges of e-mails, some twenty years ago, between Diane Strong, a diver and writer who chronicles her diving adventures in media around the world, and Bill Curtsinger, a photographer for National Geographic who is part of this story. Over the years since, I have read his byline and seen his photographs from the far reaches of the Earth.

I found myself remembering some things as though I was living them again, as though no time at all had elapsed. It was a photographic memory, and I don't have a photographic memory-far from it. I believe that the *ghosts* of Micronesia may have come for me and taken possession of my mind, guiding my fingers on the keys.

The story could be fiction for all I know; fiction only because certain parts were then and are yet ambiguous. It is a story meant to note or clarify a number of small, real, life-changing events as I lived and remember them: events that occurred in a place an ocean away, in a time far from now. I was there. The people are real. Some, like a rare few, got to lead fictional lives once-in-a-lifetime, Beyond Time.

My story begins one morning when I wandered into the Maramar Hotel, brooding, and found Carlos. It was ten o'clock in the morning on a weekday. I sat down with him. We shared some *poraus* (news) and had breakfast. He told me about his upcoming gig, for the Curtis's as captain of *Tri Amora* for a National Geographic expedition to the Western Islands of Truk[2] and Eastern Yap. So, Carlos had a new job. My Peace Corps work was winding down. I was aching with jealousy. I longed to be again in the mystical Western Islands. It was a spiritual thing, and besides, I desperately wanted to leave Truk. Finally, he said, "Do you want to go?"

The Maramar Hotel

When he was around, Carlos was a fixture at the Maramar Hotel: the place Russ and Verna Curtis built. It was named after the floral leis worn by Micronesians. Russ and Verna were a Quaker couple who had been out there for years and raised two girls who were then attending college in Hawaii. The Maramar had a screened restaurant and bar open to the constant trade winds and looking out on a broad green lawn punctuated with coconut palms. The yard ran down to the harbor, dotted with rusted scuttled ships, and beyond lay the island of Udot floating serenely on the lagoon.

We volunteers were all regulars at the Maramar for all festive occasions. This is where we said goodbye to Killer Miller when they finally let him out of the Truk jail on the murder rap. We had often seen him with the work gangs clearing brush beside the road with a machete, guarded by a policeman with only a stick. He left next day for Southeast Asia to fight for a paycheck with the Pathet Loa. He didn't care which side he was on.

The Maramar was where 'Shakin' Aetkin (Navy, UDT) drank and raged against the hippie, Commie Peace Corps volunteers. After drinking all night, Acken would head out into the lagoon with his dynamite to blow up live munitions still stowed in the holds of the sunken wrecks of the *Ghost Fleet*.[3] After thirty years submerged, the explosives had become unstable and dangerous.

1

BEYOND TIME

The Maramar was where LoveMe (her given Trukese name) served beers every night to the Trukese men until their eyes were red as hot coals and the table and then the floor around the table were stacked with empties on top of empties. The men would then stumble out and into a rusted-out truck and try to run the PCVs off the road.

We Peace Corps volunteers came to the Maramar for the delicious hamburgers that Russ and Verna served up.

Carlos

In his column in the Honolulu Star (Nov. 6, 1996), Charles Memminger said, among many true and wonderful things about Carlos Viti:

---∞---

"As skipper of a trimaran, he once found himself beating into a fierce wind, being driven by a storm onto a deadly reef."

---∞---

And all I can think is that we hear and see, even do many things, and then imagine what we will. Or maybe that was yet another story when he shed another of his allotted lives. I didn't know all the stories. I only knew two or three. I know that Carlos had a brush with death as a Peace Corps volunteer on an outer island when his appendix burst. A seaplane dropped into the lagoon and got him off to a hospital. After many surgeries, they saved him. He was lucky. There are no seaplanes in Truk anymore. I really don't know how many lives Carlos had left when we knew him in Truk.

Memminger also said of Carlos, that he was,

BEYOND TIME

---∞---

"(like) Errol Flynn, a fearless yet humble rogue who had cut a swath across Micronesia."

---∞---

Carlos was gone off the island a lot. We always liked to imagine that he was doing something exotic or mysterious in some strange place. When he came back to Truk, he always had that grin on his face. The grin covered the fact that Carlos was profoundly private and self-sufficient. He didn't talk much but when he did, in a soft voice, you wanted to listen. He didn't mix much with the volunteers and, for us, he was hard to know. I thought of him as very deep, and very Trukese, for a White man. He could have attitudes. There is a wonderful photo of him, grinning, at the helm of *Tri Amora*. Behind him in the pilot house window is a poster, in the psychedelic style of the day. It says, "Tourists Fuck Off," expressing the other side of Carlos. He felt that tourists had no place here.

Not that many tourists came to Truk in those days. The only one I remember was Bobby Darin ("... scarlet billows start to spread, hey...."). Somebody said this is where he wanted to come before he died. I saw him on the beach one day in Neiwe as I was on the path home, crossing the polluted creek that drained the *taro*[4] swamp. His companion was a beautiful young woman with exaggerated breasts in a yellow bikini. He was very pale, dressed in trousers, long sleeved shirt and beret; very thin--cancer.

Of all the choices he had left, I wondered why he wanted to come here. The forest on my right was full of *yat* (young boys) in hiding, peeking (*nikich*). Exposure of a woman's thigh is very taboo here. The most erotic word in the Trukese language is *angerap* (tuna): a woman's thighs, firm and taut, like the powerful flanks of the tuna fish. What untold fantasies the legs of Bobby's companion were arousing in the boys. I was the only one looking at her proud and costly tits.

Carlos had gone to school with Cheryl and me at UC Santa Barbara in the late 60s. We had mutual friends but didn't know each other. By near the end of my Peace Corps career in Truk, I knew Carlos only a little better. I knew he was courageous and nearly fearless.

I first saw what he was made of one night out on the western island of Puluwat while on a Truk government field trip with him. He had a big role of Dacron with him on the trip out. He wouldn't tell me what he was going to do with it, but I knew that it was something secretive and mysterious. In the evening when the *Truk Islander* got to Puluwat, Carlos rousted me and said, "Hey, come with me." We headed up a path through the jungle in semi-darkness. He had the roll of Dacron under his arm. "Where are we going"? I asked.

"To Hipour's[5] house."

"What are we doing?"

He just said we have a *"mwich* (meeting)."

When we got to Hipour's place, a traditional hut of woven pandanus and a palm thatched roof, it was after dark. As soon as I ducked under the heavy breadfruit wood lintel of the doorway, I was struck dumb by the sight. Inside the hut was a perfect vinyl tile floor, in red and blue squares, perfectly laid, polished to perfection and seemingly luminous in the dim yellow light of a kerosene lamp. I thought I had stepped into the Star Wars Bar, out of place and time in this wet, twisted tangle of jungle lost somewhere in the Western Islands. The effect on me was dizzying. Seated all around on the floor in the semi-darkness were Hipour and all the other Puluwat navigators. Ikuliman was there. In the dim light of a lantern, colors: red, yellow, and blue transformed the breathless aura of the room; the light reflected, flickering, from the men's glistening faces and tattooed torsos and limbs as they greeted us. I was freaking out. "Why do these guys want to talk to us?"

I remember, after some food and some joshing around, Carlos was seated, cross- legged in front of Hipour, who appeared to be master of ceremonies. Carlos reached out and gave him the roll of Dacron.

It was enough to make a new suit of sail for every canoe on Puluwat, and it was the only thing they needed that could not be obtained right here on this coral spit or in the sea around it. It is highly prized and the only thing that has changed on traditional voyaging canoes in twelve hundred years.

Finally, Carlos revealed his scheme. He spoke in Trukese and I understood enough to realize that, to my shock, he was literally daring all of them to go along with it. Suddenly, you could have heard a pin drop on that vinyl floor. Now I was really freaking. I think he may have used up one of his lives right here. I felt as if I had. Hipour was a bear of a man. His arms were massive, like the limbs of the breadfruit tree; a tattoo of the rising sun covered each shoulder. It sounded to me like Carlos was insulting the man, one of the most respected in Micronesia. There was a long pause in time; an eternity was passing. Nobody moved while Carlos spoke. Carlos and Hipour looked unblinking into each other's faces. All were silent as Hipour listened. Carlos told him he could have the Dacron, but there was a condition attached; Hipour must sail one of the Puluwat canoes to Guam; a voyage of 550 miles that had not been made in a traditional canoe within living memory.

Hipour scoffed at Carlos saying, "I have been to the Marianas."

Carlos chided him and said, "Yes, in the Queen Mary" or words to that effect[6]. That did it. The silence was broken. There began a murmuring. I was dying a thousand deaths. I looked at Carlos who of course was now grinning. Notwithstanding the Dacron being held hostage, I sensed in Hipour there was a modicum of respect toward Carlos, enough that Hipour would not risk losing face to this *Haole* whom he would otherwise have disregarded. Then Hipour looked square at Carlos and said quietly, gravely by contrast, "Fine, but you will come with us."

Touché! Now Carlos had to save face. And so, it was done. Carlos was going to be part of the history of Micronesian voyaging.

Such a trip, from Truk to Guam, in a traditional canoe, using traditional navigation, had not been made in the 20th century. But Hipour knew the way, and Carlos knew he did. He had even proven his knowledge when, a couple of years earlier, he had navigated David Lewis' modern steel ketch, *Isbjorn*, to Saipan. The compass and modern implements of navigation were stowed on the voyage. Hipour employed only the traditional sailing directions, which although unused in living memory, had continued to be passed down from generation to generation in the respected fraternities of Puluwat navigators. Ikuliman, near eclipse in old age, also knew the way and was chosen to navigate a traditional voyaging canoe to Guam for the first time in a hundred years, an event to which Carlos would be witness. It was sometime in 1972.

This was exactly how a renaissance in traditional voyaging began in Micronesia in the last quarter of the 20th century; a tradition left fallow, a cultural heritage kept alive only in the minds of a handful of men but close, very close, to vanishing from this earth.

The following morning, after a breakfast of rice and fish, which we ate from a large communal bowl with Hipour's family, I left Carlos on

Puluwat where he would prepare for the journey. The *Truk Islander* was riding quietly at anchor in the lagoon. Carlos and I walked to the beach where a skiff was waiting. Damn! I wanted to go too, but I went on to Pulusuk Island with a load of construction materials in the hold, about to have a personally formative experience of my own.

My own epic voyage with Carlos would come about a year later when David Lewis returned to Truk in 1973.

We didn't see Carlos for a while at the Maramar. When he did return, however, the photographs he took and the story of that voyage he told were remarkable. One hundred years of vacant history (only vacant in the Western mind anyway) came alive. It was farfetched, improbable; an all-time RatFuck.

> Def: *RatFuck (RF) (v)*; the serious pursuit of a generally unworthy, or inconsequential cause or activity, the outcome of which cannot be predicted. *(n.)*; a non-trivial, purposeful, yet pointless disturbance of the status quo- **The Peace Corps.**

> Def: *RatFucker (RFer) (n.)*; a person who engages in *RatFucking (RFing) (n. or v.)*. A Romantic. The Trickster- **Carlos Viti.**

The whole thing was that Carlos had a friend in Guam who owned a motorcycle shop. He told Carlos he wanted to have a traditional voyaging canoe to display in his shop and, of course, Carlos said, "I'll get you one," without blinking I'm sure.

The trip was long and arduous but there were amusing moments I know Carlos enjoyed. Along the way, when a plane was seen above them the crew would force Carlos to lie in the bottom of the canoe covered with a pandanis mat. They thought if a White man were spotted, he would be mistaken for a "spy" and Navy gunships would be on them in a minute. Finally, they slipped into the U.S. Naval Base in

Guam under cover of darkness. Hidden amidst the U.S. Pacific Fleet until morning, they presented themselves to customs: five guys from the Stone Age and one scrawny white guy in a sailing canoe, only to be told they could not land without passports. No papers. Right! In the end the guy did get the canoe, and everybody "self-deported." The crew was happy, and they all had a good time eating ice cream and riding back to Truk on the Air Mike jet in *thu*, the traditional loin cloth. What an RF!

When the rival Satawal navigators heard about the Puluwat navigator's accomplishment, they had to do him one better, of course. So, they went to Guam and further to Saipan, *and back*. Soon many such trips were being made, back and forth, from the islands all over the Carolines, so as to become routine. It was a renaissance in traditional voyaging.

In 1976, the Carolinian Navigators became rock stars. Mike McCoy brought the great navigator, Piailug, from Satawal to Hawaii and Piailug steered the newly built Hawaiian double-hulled canoe, Hokule'a, to Tahiti, as his ancestors had done a thousand years ago. It was for a celebration of the American Bicentennial, not a celebration of the historic migration of a people[7]. The whole Hokule'a thing was controversial. Whose heritage was this? I get it; to get money to do it, from the politicians, it had to be about "America." Some were unhappy about it. Piailug, in his navigator's grace and stoicism, remained above it, and stayed the course. The Hawaiians mutinied. Piailug made a precise landfall on Tahiti after a sailing distance of 2,500 miles. Movies were made of it, and books written. The whole affair became a reality TV show.

Pulusuk

As Carlos, Ikuliman, and the others began preparations for their historic journey to Guam, I was miles away from Puluwat on the neighboring island of Pulusuk. I carefully watched the unloading of the *Truk Islander*'s cargo; all of the materials for the dispensary that I had estimated from the plan that I had drawn, and now intended to build. On an earlier trip to the Mortlock Islands lying east of Truk, I had stood by, helpless, as family of officials in Truk offloaded all of my materials before the intended destination. It wasn't going to happen this time. In a perversion of the natural order in Truk, I got to decide who got the stuff.

The cargo boom of the *Truk Islander* swung out over the side with each load of cinder block, bags of cement, rebar, timber, and corrugated iron roofing. As the small freighter rode a tall rolling swell, the skiff alongside would rise to meet the swinging pallet. At just the right moment, the hook was released, and the skiff with its load would plunge into the next trough. It was a mesmerizing ballet under a blazing sky. It took many rounds of this dance, and many trips in the skiff through the shallow channel in the reef, to get it all to the beach. I rode in on the last trip with a jar of peanut butter and a bolt of brilliant blue cloth, out of which I intended to fashion my own *thu*[8]. I would stay until the ship returned. When? none could say. We were Beyond Time.

I had been shown my quarters, a bachelor hut that I was to share with Santy, when Inocenti, who said he was my "friend for life," came by, took me by the hand and led me out into the forest to meet the rest of the guys. I wore my *thu* for the first time and fit right in. We drank *tuba* (palm wine), spiked with yeast, foaming and bubbling from a ship biscuit tin. A half coconut shell of the milky brew was passed around a circle of us squatting on the ground. We drank, told jokes and stories, passed shared cigarettes, and laughed until I couldn't get up off the floor for two days.

Michael's family was to care for me during my indefinite stay on Pulusuk. Michael was trained by a former volunteer, Marvin Kretchmer, in the use of cement, Western building materials and techniques, and had organized a crew for the dispensary construction with the materials I brought with me.

The place we were given to build was on the main pathway through the village near the intersection of the path to the landing--practically "downtown." The dispensary would find some use in the village, I was sure. When construction neared completion and the roof was put on, the island girls appropriated the shaded porch bordering the path. There would be giggles and gossip as the island boys paraded by.

The health aide, like a fat angel, hovered over the project. He was a big, happy guy with a front tooth capped all in gold except for a small bit of brilliant white enamel showing, in the shape of a heart. He was never without a smile on his face. He was the only one on the island, besides the teacher who had a paying job. When the dispensary was completed, the hospital in Truk would send out a kerosene-powered refrigerator for his use in the new facility. Everyone was excited about the prospect of cold beer. RF!

BEYOND TIME

*"Well, I'm going downtown with my Gold Tooth displayed
I got a fine fox in front and two more in the back ...."*
-ZZ Top-

For my first few days on the island, they assigned a six-year-old boy to follow me around so I wouldn't get into trouble. That should give you an idea of my own functional age in this place. There are many things I must learn from a six-year-old, in order to survive. The little kid explains the *"benjo"* (place where one relieves oneself- a Japanese word). In fact, it was just a log at water's edge. This was where I was to go at night to relieve myself. The tide takes away the night soil before morning. At all other times, this place is taboo. When Western visitors come to Pulusuk, however, they invariably seat themselves, after debarking the skiff, on the *benjo* log which is conveniently located on the beach near the landing. Now that's entertainment to the screaming delight of the children! I'm in on the secret.

Santy, the man I lived with on Pulusuk was a serious man. He was perhaps forty some years old. Old enough to be taken seriously in the culture. He was a navigator, and master canoe builder. He must have had polio as a child since one leg was withered and useless. He used a sturdy stick under his arm in its place. It made no matter, as he was able as any man. Aboard the canoe, when we went out fishing, he swung himself about the boat with ease using his strong hands, arms, and huge shoulders.

We trolled for tuna using modern jigs with hand lines, at six to seven knots. I caught something I couldn't hold onto. The monofilament

line cut deep into my hands and I had to be helped with my catch. It was a five-foot shark. It was brought aboard, clubbed to death, and returned to the sea. The Micronesians don't eat the meat of the shark. Some of the tuna that we caught, we had for lunch, seared on a fire of dried coconut husk in a white enameled pan set out on the outrigger platform.

On occasional afternoons, Okapi and I took a small paddle canoe out to the drop-off. As he spear-fished along the edge of the reef, I would swim at the surface and gaze down into the awesome vastness of the depths along the wall. Okapi carried just a spear fashioned from old torpedo netting, with a natural rubber sling. He wore handmade goggles, carved of wood and plastic, and fashioned a pair for me. After several fish were in the canoe, the shark shows up.

Okapi would spear a fish at twenty or thirty feet below, turn his back to the wall and bolt for the surface, arm reaching upward with the fish on the end of the spear. Out of the depths, the shark would come, going after the impaled fish. In the last ten feet, Okapi and the shark are belly to belly less than an arm's length apart, racing to the top until the fish is thrust above the surface and the shark peels away, and disappears back into blue infinity. After three or four rounds of this game, Okapi finally says, "enough" and we swim the small paddle canoe back to the channel, and onto the beach. Back on the beach, Okapi's catch was spread out on a woven mat and then divided among several families, a practice of communal living since time immemorial.

When I first arrived at Pulusuk and took up residence with Santy, he was laboring on a small working model of a voyaging canoe. It was precise to the last detail, even to *yolool* (coconut fiber sennit) lashings that tie together the irregular planks of breadfruit wood to form the hull. As I watched him fashion the model, I was also observing his work on the real thing, almost 30 feet long, in the canoe house. I could see that the proportions, materials, and details were identical. I asked if I might have it when he finished. He said it was mine. On the last day, he was preparing to paint it and I asked him not to. Santy seemed disturbed that it wasn't finished properly, but I wanted it to look just as the real thing that I had watched rise in stately grace from the floor of the canoe house[9]. I didn't know how to tell him that, as an architect, I would never paint a piece of wood. I paid Santy as much as I could, and he gave it to me unfinished, in its natural beauty.

The Way Of Wood

I can't remember how many *wuut* (canoe houses) lined the leeward beach of Pulusuk Island, but one, I can never forget. Not that it was different in appearance from the others, but rather for the work going on there. The *wuut* itself was a masterpiece of work. In each corner were placed upright, deeply buried, trunks of the massive breadfruit tree. These were the girth of three men, cut at a branching Y to make a cradle for the only slightly less massive beams reaching up to bear the roof canopy. The roof of the *wuut* was a thatch of coconut fronds, splayed just as they hung from the palm, and laid up from the eaves, in weatherboard fashion, to shed the rains. The gable ends were enclosed with woven mats down to a level that one must stoop to enter. The eaves of the roof were at waist level and soared to the massive ridge beam at a height of some twenty feet.

Pains were always taken to see that any piece of natural material used for construction was always situated in the same relationship to earth as it had been growing in nature.

Construction of the dispensary was no exception. Spirits, any spirits, are not to be messed with (wouldn't your spirit be upset if you were planted in the ground upside down?). This was true, as well, for the spirits of the ubiquitous, sawn Oregon Douglas Fir that I brought with me as construction material. Before coming to me, it had been dragged coarsely without reverence of any kind from the dripping forest of the American Northwest, ripped with a saw, trucked without ceremony to a grinding West Coast port, dumped in a pile in Guam, and beaten by the weather before finding its way to Truk.

My 2x4s were finally respected here on Pulusuk island, in what was to be their final Eden and place of rest. Before we placed any timber back into the earth, we spoke of its history and tried to understand how it had grown. It must be placed as it had grown. Sometimes it was hard, but we always paid our respects before anything else went forward. If there was any uncertainty about a particular 2x4, it was set aside and conjured for a day or two until a consensus arose as to how it had grown.

I was trying to read Ken Kesey's *Sometimes a Great Notion*, the brooding, dark story of an Oregon logging family set by the Rogue River. I just couldn't get into that place so distant from here. I felt that I was no help in knowing how the 2x4 once grew in that misty gorge. My parochial Western mind could not encompass such a range.

I usually ate the mid-day meal with Santy and the other craftsmen in the canoe house. Women left our food just outside. We ate from a common bowl with one hand, me the only one bothering to wave at the flies with the other. Some of my crew were also working on the new voyaging canoe, so the noon meal ended the work of the dispensary for the day. Their real work loomed above us in the semi-darkness of the vast thatched roof, each end of the massive keel resting on a stump of a coconut tree. It was almost thirty feet long and was nearing completion.

Perfect Evolution

I spent as much time as I could in the canoe house. The floor was a damp, fragrant mattress of shavings cut from the seasoned heart of the breadfruit tree, as planks of the hull took shape. Other kinds of wood, each chosen for characteristics suited for its function, were employed for other parts of the boat. The outrigger beams that supported the *tam* (outrigger) were of the more supple wood of the mangrove, which could flex, but not break, under the strain that would be put upon them. The *tam* itself was a piece of dense, solid breadfruit meant to be a counterweight to the overturning vector force of the wind on the sail. The canoe was a true *Proa*[10] of ancient design lineage being built in front of my eyes as it had been done for a thousand years[11].

Here on Pulusuk, a great breadfruit tree from the forest was being transformed into a sleek and handsome seagoing craft. Perhaps this was a tree felled in the forest by a typhoon (perhaps the one before Amy). Living breadfruit trees are never cut. Chip by chip with adze and machete, the irregular planks were shaped to fit into the asymmetrical yet hydro-dynamically correct form of the emerging hull.

I believe that the Carolinian Voyaging Canoe is a perfect evolution. With the materials available to the Micronesian canoe builders, you cannot make it any better in terms of function and performance. Sure, you could make it out of different materials, but the engineering principles employed would require the same balance of form, and the same balance of flexural strength, rigidity and mass. A modern proa constructed of molybdenum unobtainium can exceed 55 knots to windward in 24 knots of wind. The reason it can sail faster than the wind is that, as it accelerates, it makes its own wind. If we could ignore the Newtonian force of drag, the theoretical speed of a sailboat to windward is infinite, and without any power other than the wind.

Magic and Awareness

The traditional canoe house is the reserve of men: mariners, canoe builders and the navigators. Women may approach but cannot enter the canoe house. Despite the introduction of Christianity and the Church's efforts to suppress it, there is still much magic in Truk. The magic of women and the magic of sailing are not compatible. The canoe house shelters the canoe as it is constructed, or under repair. As the meeting place of sailing men, its magic prevails and must not be tainted. What is it about the magic of a woman? Is it simply intimidating? The power of men pales before it.

Magic is palpable everywhere in Truk. We had medicine magic hanging in the rafters of our house in Neiwe[12], in a small round pouch of tightly wound coconut sheaths with knotted leaves tied to it, enforcing a restriction that no person should touch or trespass against this medicine.

Traditional navigators had abandoned some aspects of traditional magic and taboos, practical sailors that they are, and especially since Hipour deliberately abandoned them and made a successful voyage. Nevertheless, It certainly still seems central to the practice of navigation, which is itself central to the culture, but what is it?

I really think magic is the creation of an *absence* of everything that separates us from the natural world that distracts us from a true awareness of how we are connected to everything. To a Micronesian navigator the vast panoply of sea and sky is full of signs. There are signs everywhere – if you know how to look, know how to feel, how to hear, how to smell. It is not magic to be able to see it all. It is simply knowledge and awareness: an awareness we in the modern world are blind to.

Carlos told me about magic practiced on his trip to Guam with Ikuliman. I believe it boiled down to a way to spend some time in the moment, open to awareness, out of which came a consensus and alignment with nature about when it was best to sail for assurance of a safe passage.

On this voyage, Ikuliman had chosen to first go to Pikelot, a waypoint on the journey to Guam. Before embarking for the remainder of the trip, the mariners sat cross-legged on the beach, each next to a pile of coconut leaves. In great concentration and contemplation, they would carefully peel each leaf along the veins and set aside the strips of coconut.[13] This appears to me to be a process of focusing the senses on the Now, and full awareness of one's surroundings: the sea and sky, and the signs that are there, which we Westerners cannot see and cannot contemplate.

I believe it was a practice of being in the "Now." Under the spell of magic came awareness of all the signs in the sea and sky that can be known. By this means, they chose the day of leaving.

Stranger In A Strange Land

There on Pulusuk, all things seemed in balance. Always, there were many tasks underway about the island. In the canoe house and at the dispensary site, children, even toddlers, sat next to their fathers, toying with the razor-sharp blade of the adze and machete, becoming familiar with tools that their survival would depend on. There was no hurry, but there was no idleness anywhere to be found. There was industry, but no stress. In a day, there was time enough to shape a voyaging canoe, mend a net, spearfish on the great wall of the Pulusuk reef, or gather breadfruit from the forest; to plant *taro* in the marshy interior, prepare food, tend a child, or bathe in the forest beside the clear freshwater lake.

The fragile "lens" of freshwater makes life possible on these islands. It is said the great limestone pillar of Pulusuk, formed from accretion of corals over a billion years, is the shape of a mushroom (actually wider at the top than at the bottom) as it rises from the sea bottom thousands of feet below, its texture that of a sponge. Riding on top of the denser saltwater invading from below, is a thin layer of lighter fresh rainwater. Atop this porous mushroom, overlooking the great Marianas Trench[14], the deepest point on earth, survives life - Beyond Time.

For but a brief time, I was immersed in this place beyond my imagination, strange to me as I was strange to it. I don't know what it would take to be really a part of both worlds. Maybe Carlos knew how. I was there only long enough to see how such a life *could* be lived; to see the peace and the beauty in it, the rhythms driven by nature. Other times, I wanted to cry for something, I don't know what, or just to be home instead of a stranger.

On Sunday afternoons I would go out into the forest in the interior of the island. Magnificent breadfruit trees towered above me. The rays of sun fell through the radiant green canopy and lit up the brilliant ferns at my feet. It was like a cathedral. The roar of the surf out on the reef was muted. I sat back against an ancient breadfruit tree and read Tolkien. Once, I went to the far, windward side of the island where the surf booms on the reef. Along the vacant beach, I tried to conjure *chernikin*, (the Sea Ghost). Some of the volunteers on outer islands had felt its presence. I waited in fear and anticipation, but it never showed itself to me. Lucky for me me since all ghosts are malevolent, and *chernikin* especially so. It would have taken strong medicine to save me.

Ghosts

Ghosts are everywhere in Truk.

People in Truk, if they survive childhood, live long lives. The graves that dot the family compounds are mostly small ones, the children. They are decorated with sparkling broken glass and other colorful objects. When and if a family can afford batteries, a small transistor radio, under the little cross, plays music and *poraus* from Radio Truk. This is to keep the ghost of the child occupied so that it plays around the gravesite and does not go out to make mischief in the village. When we didn't hear the radio, Cheryl and I would bring home batteries from Truk Trading Company to our family. It was always comforting to hear the soft melodic sound of slack key guitar and singing voices from the jungle behind our house. No use taking chances with ghosts.

Ghosts can take possession of you and cause you to do things not of your will. Sadly, drunkenness is seen as possession by some such ghosts. It can cause you to do things you would not do. Trukese men drink until drunk and whatever they do under the influence of the ghost is forgiven, even murder. Christianity, brought to the islands by the Jesuits, only reinforces traditional beliefs. I believe this is why it is widely accepted, with its emphasis on living spirits i.e. the Trinity, and forgiveness. Christianity even gave a name to the place where spirits of ancestors dwell- *purgatory*. It was on Pulusuk, in Father Fucky's church, that I first felt the stirrings of my own later conversion to the faith.

Father Fucky

You might say that my conversion to Catholicism later in life had its origins in my stay on Pulusuk Island. On Sunday morning, every living being on the Island went to Church. There were only two: Protestant and Catholic. There are no non-conformists here, and I wasn't going to be the first. So, I went to Church on Sunday morning with everyone else on the island. The choice of denominations was determined for me by the clothes I had not brought with me.

At the small Protestant Church, men were required to wear *rousers* (trousers), and women must cover their breasts. Well, I hadn't brought any trousers with me, and not too many other men had them either. The women weren't inclined to want to cover their breasts, so a few men with *rousers* (trousers) attended the Protestant Church and the Catholic Church was the place to be on Sunday mornings!

It wasn't so bad. All the pretty girls were there, fresh and dressed for Church with a *maramar* (lei of flowers) in their hair. The fragrance of the flowers, and the scent of fresh coconut oil in their hair and on their bodies was intoxicating. We sat on the white coral sand floor of the Church. I didn't speak Trukese well enough (especially the dialect of the Westerns) to understand a single homily, but the ritual was comforting.

FATHER FUCKY

The flock of faithful was watched over by Father Fucky, a Jesuit priest who had been out there since 1948. Yes, that is what everyone called him. His real name: a good Irish name, Fahey, was unfortunately an obscenity in the local language. So, to avoid embarrassment and yet address him properly, everyone simply spoke the Father's name in its English translation of the Micronesian word, and he was called Father Fucky. A respectful solution, I thought, to a difficult cultural conundrum.

Rivalry

There are rivalries in Micronesia. I have spoken about the rivalries among navigators which drove the renaissance in voyaging. There were rivalries among the Christian missionaries as well. I had a brief but memorable encounter with another Jesuit, Father Rively, on Lukunor. Like Father Fahey, Rively had been out in the islands since the war. Since my building materials were hijacked before I arrived, I had nothing to unload at Lukunor except a few useless, weathered 2x4s. Sensing the anger my paltry pile engendered in the islanders, I felt foolish. Licking my wounds, and now with down-time, I sought out Father Rively's house. Down a path through the trees, I found it in in a dappled grove of palms, a house stately–constructed of cut coral stone, glittering in its bright lime wash, looking Victorian with its steep gabled roof. Anchored in the lagoon, within sight of the high, broad porch, was his beloved ancient wooden Tahiti ketch, *Star of the Sea*.

As I approached, he came down to meet me with a warm gracious smile. I knew I was going to like him. He invited me in, where I found his living room piled with yellowed and tattered yachting magazines all over the furniture, and all over the floor. We adjourned to the veranda where, eager to talk (and being a good Jesuit, a bit of wine was involved), he regaled me with colorful stories. Both of us being sailors, the conversation turned to boats and sailing, and his exploits on the *Star of the Sea* sailing down to Rabaul, New Guinea, for periodic refits.

Apparently, there was a fine rivalry between the good Father and Herr Mueller, the young Protestant pastor at the other end of the lagoon (maybe the island of Oneop). Herr Mueller had a slick, imported speedboat and occasionally liked to disrupt Father's Mass. He would water ski right in front of the Catholic Church on Sunday mornings and of course all of the congregation would run to the beach for the spectacle, hooting and hollering. Father Rivley would retaliate each July 4th and sail the old "Star" across the lagoon with a load of fireworks which he would proceed to light off. This of course had the similar effect of ending Herr Mueller's sermon as his Church emptied to the beach, all hooting and hollering! A good kind of rivalry I thought.

Leaving Pulusuk

After six weeks on Pulusuk, I was sick. The amoeba and worms were weakening me. I was ready to leave--a leaving that would be bittersweet. On my last day, as the *Truk Islander* returned, I paid Michael the wages he and his crew had earned. It was meant to go to him and the others who had built the dispensary, but there in my presence, Michael gave it all to a few men not known to me who had their hands out; those of higher status, but who had no connection to the project.

Everything is distributed in some order established before history began. There were winners and losers; it was all decided long ago, and the order will never change.

That night, as I leaned on the cold metal railing of the *Truk Islander*, there was only the noise of the motors, fans and mechanical equipment that made her work. My nostrils were full of the smell of diesel and red lead. The only natural thing I could see was the dark shape of the island's canopy of trees against the star-lit sky; the only thing I could feel of this place was a gentle breeze blowing off the land, all of its sweet fragrance erased by the smell of grease; the evening meal just served in the Captain's mess. I wasn't hungry.

The lights of the *Truk Islander* streaked out over a gentle swell--the only electric lights within thousands of miles--save those of Truk. One night those went out as well. What an all-time great RF that was when somebody fell asleep and a thousand gallons of seawater, instead of diesel fuel, were pumped into the tanks feeding the big Caterpillar power generators and blowing them up! The tanker's pumps were used to pump seawater into the ship's tanks which lifted the oil, being lighter than water, out of the ship's tanks and to the storage tanks for the generators. Someone had to be awake (or sober enough) to shut off the seawater valve at the right time. But I digress.

Like Satawal to the west, Pulusuk has no lagoon. We were anchored in an open roadstead in the lee of the island, along the edge of the fringing reef. The bow hook of the *Truk Islander* was set on the edge of the reef in but three feet of water, and there were thousands of feet of water below the keel! It must be the world's most awesome drop-off. I remembered the wonderful days with Okapi spear fishing along that vast wall which appears to simply drop into eternity.

Leaving Pulusuk was bittersweet.

Water World

Satawal and Lamotrek were administratively in Yap District, Puluwat and Pulusuk in Truk District, simply political distinctions. The districts were established when Micronesia was the Trust Territory of Pacific Islands under American rule. Names have changed since independence but nothing else has changed in the Western Islands. Political boundaries are as meaningless now as then, just lines on a map. The atolls that are spread for hundreds of miles between the great lagoon of Truk and the mountainous archipelagos of Yap and Palau have been, and are, a world unto themselves. Affiliation is between family and clan, not to some sovereignty. Being between the atolls of Puluwat and Satawal, we were in the center of this Water World.

There is no connection between Yap and Truk by government ship. The people of the islands are linked, as they have been for thousands of years by their ability to construct competent ocean voyaging canoes and navigate across the trackless ocean between themselves. Is it any wonder that it is the navigators (palu) who are kings of this domain with status higher even than chiefs?

Tri Amora, for a short time and in a small way, was pressed into this traditional connection. She carried a family from Puluwat to reunite with family in Satawal and brought Manipi, Flag Chief of Puluwat back with us from Lamotrek to his home island. She carried Mike McCoy from Truk to Satawal to re-unite with his adopted clan and carried bags of cement from Puluwat to Pulap for Father Fucky's church.

She carried David Lewis from Truk, to learn the secrets of the navigation. The irony is that, as this knowledge is spread to the West by his work, it is in danger of being lost where it began.

Tri Amora

———∞———

"We (writer David Lewis and myself [sic]) just put out a request for a boat that had the range and the capacity to accomplish our assignment."
-E-mail from Bill Curtsinger to Diane Strong-

———∞———

Russ, with local help, built the boat: a trimaran, of Piver design, from a kit of pieces over in Yap. He named her *Tri Amora*[15]. She was 44 feet in length, of plywood construction. With her ketch rig, pilot house, shallow draft, and off wind performance, she was an ideal boat for skirting the reefs and shoals of Truk Lagoon, and her broad deck was perfect for diving expeditions on the wrecks of the *Ghost Fleet*, trips we made on many weekends. She was a complement of form and function.

For all of her virtues sailing off the wind, *Tri Amora* did not go to weather well. She was not close-winded and made significant leeway when sailed up-wind, complicating navigation when tacking to windward.

In many respects, she was like the boats the Micronesians themselves build and sail. In fact, they taught us how to build this type of craft. After all, sail was invented only twice in the world, once here in the Indo Pacific with multi-hull craft, and once in the West with single hull keel boats. The two approaches are very different. Modern boats of both types are built, and each has its devotees.

In the summer of 1973, Russ had chartered *Tri Amora* to National Geographic for an expedition to the Western Islands. He hired Carlos and Tomoichi to crew her. I was a thankful add-on. Tomoichi was an earnest young local man. He would turn out to be an invaluable hand and faithful companion through trials that were to come for this expedition.

Rock Stars

The day of leaving was hectic. At the dock, Carlos and Tomoichi were still installing the new diesel that Russ had ordered from Japan. I presume he used some money National Geographic had advanced so that our expedition would have "a boat that had the range and capacity to accomplish [its] assignment."

About noon, Carlos and Tomoichi were still up to their elbows in grease and oil in the bilge when Dr. David Lewis and his entourage; a photographer, and a girl, arrived like Rock Stars in the back of a Datsun pick-up with their *pisek* (stuff). My first thought was, "He doesn't look like a Rock Star, but hey, he's got a roadie and a groupie."

I was looking forward to sailing with Lewis, whose legend as a master navigator and mariner had preceded him to this far place; this place where none can remember it but as anything other than a sewer outfall of American culture.

The U.S. Post Office was the only thing in Truk that worked on a regular basis. It was the de-facto Sears and Roebuck Catalogue outlet, piled with pisek that arrived regularly on the Air Micronesia jet, ever replacing what stops working or rusts away, out here where rust never sleeps. Here is where we bestow on this, our last colony, all kinds of useless stuff to feed the rust, hoping the people will love us more than the Koreans who want their fish, or the Japanese who simply want their dead[16].

I was looking forward to getting back out to the Western Islands, beyond the stain, Beyond Time, one more time, before I went back to America.

Before Lewis even introduced himself (I'm sure he presumed we knew who he was, and he knew who Carlos was, and this is all that was necessary), he explained to us that he would take no part in the navigation or operation of the vessel. This was to be in the capable hands of Carlos and his crew. I looked at Carlos, and it was all confirmed. He didn't flinch. Lewis was to stay true to his word. We were on our own.

Carlos measured each day in grand sweeping terms like the Micronesian sky, not in tiny, measured steps as to cross a mountain stream. I knew he had no practice with the sextant, an instrument of precision in a precise Western world. That was not his style. I knew however, that wherever Carlos was going, and however we were going to get there, if it was to the end of the Earth, I was going too. We were about to leave for some precious, unforgettable time where I was going to be in Carlos' world, Beyond Time. What an RF this was going to be.

Bill Curtsinger, the National Geographic photographer, looked clearly displeased with the state of things, like he expected something more "suitable." He looked displeased much of the time. Perhaps it was his experience as a Navy Seal that set him apart. His standards were obviously extremely high, and he had little time for those who didn't measure up. He was still on military time and seemed otherwise only concerned with his roadie stuff: aluminum cases of cameras, equipment, and batteries.

BEYOND TIME

The only thing I know that ever humbled Bill was a Grey Reef shark with a brain the size of a pea.

Ann Valentine (the groupie?) came aboard. She was a Pan Am stewardess in red boots, white shorts, blue bandana around her neck, and carried only a little red and white plastic Pan-Am bag. At least, I thought, she knew how to travel light. She was cheerful. We learned that she was always cheerful, except when she was seasick, which would be most of the time.

A few minutes later Mike McCoy arrived. We were going to take him back to his home on the island of Satawal, over in Yap District. He was grateful for the ride. It's the kind of place *you can't get to from here*. Mike walked out of the mid-day heat, dust, and traffic of downtown Mwuan Village. He was tall, and looked larger than life, his hair cut short in regal out-island style. He was wearing only a traditional *thu* and carried a small men's traditional pandanis bag over his shoulder. He had the Dolphins of the navigator tattooed up and down both legs and walked with the grace and dignity of those men I had seen only in one other place- the Western Islands. He was totally unselfconscious, naked, and White. He and Carlos seemed to know each other well. In contrast to Carlos, Mike liked to chatter.

Provisioning

"The National Geographic' medical department gave their field people access to a very well equipped first aid kit and any and all appropriate prescription drugs one might need in the middle of nowhere for any possible medical emergency (pain killers, antibiotics, etc.). I never went anywhere without a full complement of field dressings, prescription drugs and antibiotic ointments/pills and antiseptic solutions."
-E-mail from Bill Curtsinger to Diane Strong-

Mike went off to pay respects to some of his adopted clan from Satawal, now living in Moen. Lewis asked only one thing. Where could he obtain provisions? We sent him to a Quonset hut in the center of town next to the Peace Corps office. At the Truk Trading Company you could buy a hand-crank pasta machine for two dollars, it had been on the shelf since 1949. You could find canned lettuce. We never opened it because we couldn't imagine that the beautiful head of lettuce depicted on the stained label could be inside the can. We didn't want to know. Little else could be obtained there.

BEYOND TIME

I didn't know we had all those good drugs aboard that Bill mentioned. I never gave a thought to needing that kind of stuff. I wondered most about what we were going to eat. A couple of hours later Lewis returned with the following:

    1 50 lb. sack of white rice
    2 cases of canned mackerel (I must go to Petco now for this stuff when I get a craving)
    1 case of Kim Chi (low fat, barely edible)
    1 tin of ships' biscuits (good with peanut butter)
    1 bag of fresh onions
    1 case of Spam (nice breakfast change-up which I learned to love)
    2 bottles of ketchup (nice when fried up with the rice and onion)
    2 cases of Johnny Walker Red (not for us)
    1 case of assorted California wines (not for us)

Lewis was agitated and insisted it was time to get going. I left Carlos and Tomoichi to finish the engine and dashed over to Truk Trading Company for the last jar of peanut butter on the shelf. I grabbed a bottle of cooking oil too. It looked like I was going to be the cook. I took a taxi back to the house in Neiwe. Cheryl and the scooter were gone, probably at work in the District Finance Office with Yosua. I took a shower under the oil drum catchment we had built to collect rainwater from the tin roof. There wasn't much water in the drum. We were well into the dry season. Oh well, Cheryl would have to deal with it. I grabbed a pair of swim shorts and a long sleeved chambray shirt and took the taxi back to town just in time for the leaving.

Leaving Truk

Leaving seems to be my step-style of life. Leaving stuff behind. Sometimes ruins. It's how I got to the Peace Corps, leaving all the wedding gifts and the Saab. The ruins of the family's hope for us. "You want to go where? Why?"

We left at dusk. I remember Cheryl at the dock, atop the sea wall of recently driven sheet piles, (having missed an unexploded 500-pound bomb from the war by inches). She seemed already distant, maybe lonely (did I hope?) in her colorful sun dress and a broad white Panama hat. She waited against the darkening background of drooping coconut trees, wrecked Datsuns and the rusted tin roofs of downtown Moen. She was the only one there to see us off, the only one there with any connection to us; just standing there, no wave, just a long look between us as we drew apart.

My wife, Cheryl, and I had been through a rough patch. I had the intense feeling that this leaving was a remorseful farewell, and though I had complete confidence in Lewis and Carlos, a return seemed distant and even uncertain. In this part of the world there would be scarce safety net, nor massive reality show production support. We were on our own.

There was no wind, just the hum of the new Yanmar diesel deep under the pilothouse. The distance grew. I stood in the stern sheets, mooring

BEYOND TIME

line in hand, the last tether cast off. Was the confession to be my last memory of her?

The sky was flaming red as we struck for Udot, Tol, and the Northwest Pass. Free to leave, we would soon be Beyond Time.

It was well after dark, steering by moonlight that first evening out, when we anchored close-in at the northwest end of Tol, largest of the high islands that erupt from the center of mighty Truk Lagoon. We still had some miles to go before the pass through the reef and open ocean. We would do them tomorrow in daylight.

It turned out Mike had clan on Tol as well. They treated us like royalty, of course, especially Lewis whose legend had preceded him even here. Late into a star-crossed night, around a kerosene lamp near the beach, many stories were told in a dialect of the Western Islands. Mike translated, and legends were born again. You could plainly see that here Lewis was in his element. I was enchanted. Carlos was out somewhere on the island, getting laid. When I returned late to the boat with Tomoichi, Ann had already turned in, down in the aft cabin. Bill was in his bunk listening to music on his Sony with earphones. I asked if he would put it on the speakers. It was Elton John:

———∞———

"And Jesus, he wants to go to Venus, leaving Levon far behind…."
-Elton John-

———∞———

Sleighride

We left our anchorage at Tol in the morning, loaded down with green oranges, ripe watermelon, and some drinking coconuts, gifts from Mike's family. We left behind some of our stash of hooch, some Johnny Walker Red Label.

It was a crystal day as we came out through the Northwest Pass and turned west, running before a booming Trade. The swells were huge[17], and we slid down their faces as though on a surfboard; nineteen, twenty, twenty-one knots. We were flying! It kept up like this all day. *Tri Amora Amora* was made for this. The smiles never left our faces. We ate the cool sweet watermelon and drank coconuts out on her broad deck, it was like a picnic in motion. As the high islands of Truk dropped out of sight, I back-checked our course. I tried to memorize the picture of the place we were leaving. I wondered how long before I would see it again. We were on a crash course for Puluwat, some 150 miles distant, in some other world, Beyond Time.

Lewis never joined us on deck. He spent the day below, as he would during all of our time at sea, at the galley table scrawling madly with a pencil on yellow lined note pads - his next book, about Antarctica. I think that is why Ann was there, the ghost writer to help with that. But she was in the aft cabin, seasick and down for the first of many counts.

BEYOND TIME

I noticed the redness and dark patches on Lewis' hands and fingers. "frost bite" he said. It seemed so incongruous here at our current latitude, just seven degrees north of the equator. He had just returned from a trip, single handed, from New Zealand to Antarctica. His sailboat, *Ice Bird,* had pitch-polled off the top of a huge wave somewhere in the Southern Ocean, wiping away his rig.

Lewis told me the story of how he had met Jacques Cousteau in Antarctica. They didn't seem to hit it off very well, "That stuff about his rescuing me is a crock," he said, in his Kiwi accent, "I made it to Ross on my own. Hell, I had to ram Calypso at the dock just to wake 'em up. They were all soused to the gills!" He didn't seem to care much for sunshine either.

Puluwat

We spent a night at sea before crashing (figuratively) into Puluwat just before dawn. Shortly after anchoring in the lagoon, in the coolness of morning, there began to arise eerie, strange trilling cries from the trees. They were high-pitched human sounds but somewhere between the cry of an animal and the warble of a bird. Everywhere, there were *yat* (young boys) in the tree-tops, calling, and there seemed to be a lot of bustle on the beach. It was just daylight. I turned and looked out to the west, over the top of the reef. Against the still-dark western sky and far off, I could see two small triangular sails, golden in the morning light, against the still darkened backdrop of the western sky. They were working against the trades on port tack; it would be a while before they fetched the pass into the lagoon and home. They had been gone for some days to the island of "Pik" to harvest turtles. The arriving canoes disappeared behind a small island on the west side of the pass, and I fetched up some breakfast - peanut butter and jelly, on ships biscuit.

Pikelot, or "Pik" as it is known, like the fateful island of West Fayu that we will later come to know, are like game preserves. They are uninhabited but visited often by sailing parties in voyaging canoes from throughout the Central Carolines. They try to preserve these places for all the clans and the turtles. The sea birds are unafraid, and the sea turtle still returns at the tug of the moon to these, the beaches of their birth to brood the next generation.

BEYOND TIME

How strongly the island people feel about this refuge can be heard in the tone of anger and lament in their stories, real or cautionary, of vandals (never named) who have wasted and left behind the carcasses of sea turtle too numerous to carry home alive in their canoes to feed the community.

This morning, on the island of Puluwat, there was a lot of activity around the *wuut* (canoe houses) that lined the inner lagoon. The pungent smoke of cooking fires rose up out from the shadows of the trees and drifted out over the dark, still waters of the lagoon. There was a voyaging canoe moored next to us, tethered to one of the bent palms that reached out over the beach. Its hull was gleaming black, as though freshly painted, trimmed in brilliant yellow, a color I had seen on no other canoe. I recognized the crew as a party from the island of Pulusuk. Then I realized that this was the boat whose final construction and sea trials I had witnessed there, some months before. Then, she was still brand new in gleaming, natural, unpainted breadfruit wood plank. Having passed her trials, she was now painted. Of course, they told me she was "very fast." She looked it indeed.

The Way of Turtle Dying

When the two returning Puluwat canoes at last rounded the island and lay on the last tack to fetch the pass, it seemed the entire population of the island had turned out, filling the beach in great excitement. Only the crews of the arriving canoes, however, were to carry their prize ashore. Women came to the beach bearing bowls of food for the returning crews. Small, naked children ran among the bronzed, tattooed men in brilliant *thu* whom, like pall bearers, bore the great turtles on their shoulders, and carried the living dead through the shallow clear water, to the beach. I was soon to learn their fates.

The turtles were placed on the sand, in the shade, still living but trussed immobile on their backs and able only to weakly flail the air with a flipper that may have slipped loose. The selected one was set apart. A slit with a knife was made just below the tail. Its entrails were dragged out, and a fire was built on its carapace (belly). The intestines were taken down to the water's edge and wrung of their contents. The guts were roasted on a stick in the fire while the turtle began cooking, still alive, in its own juice. I looked into its face and there was a tear glistening in the corner of its almond eye (yes, turtles have tears). As the fire blazed up on its belly, it heaved a deep, final sigh.

Satawal

Soon we left Puluwat for Satawal, another 150 miles distant, to the west. Upon arriving in Satawal, Lewis was immediately swept into almost ceremonial greetings. There will always be respect among sailing men, rival though they may be, or centuries apart in time. Lewis was a legendary White man among Micronesian navigators. His exploits were renowned here, and theirs a passion of his. They shared the bond of sailors.

Our stay in Satawal was brief, as we intended to return after a side trip to West Fayu. We got wind that a party from Satawal was there harvesting sea turtles. Bill wanted to go immediately, hoping to photographically capture the turtles returning to their ancestral beaches to lay the eggs of the next generation. Fatefully, the return to Satawal was not to be.

Gift of Magic

Before we left Satawal, however, Lewis had traded all of our booze for all of the secrets of navigation that it could buy. We, the crew, managed to stash away one bottle of a good California Cabernet. We planned to open it when we got back to Truk, as we sailed again through the Northwest Pass. The journey to that bottle was going to be a long, hard one that we couldn't imagine.

Lewis left Satawal with a great prize (so he was made to believe). The Satawal navigators, wanting to give him a worthy gift and wanting to ensure him safe passage, gave him a small, amulet containing a pasty red-orange substance. I recognized it immediately from my stay on Pulusuk months earlier. They told Lewis it contained magic. Who's to say it didn't?

One day, toward the end of my prior stay on Pulusuk and when the fine form the new voyaging canoe was near finished, the usual lunch break turned into a feast. There were piles of fish, heaps of steaming white rice, and the freshest of pounded breadfruit, still warm and soaked in sweet coconut milk. Men and boys crowded under the huge roof of the canoe house, women and girls squatted outside and partook of their separate plentiful meal. Beiong, traditional chief of Pulusuk was present in black *thu*, stately, tall, aloof and clearly master of ceremonies.

The work on the canoe that afternoon became ceremonial, and it left a lasting impression on me--an impression of the mastery of these people who, with the simplest of implements, and the elegance of craft had come to fashion a boat of great beauty and utility from the small abundance of their surroundings.

One of the simplest and most elegant of implements was the amulet containing a turmeric of the red-orange pollen of the Frangipani flower.

The paste was smeared on a thin sennit line and the chief himself stepped forward and pulled the line taut against the hull of the canoe, from stem to stern. This was repeated until a series of orange, broken lines streaked the entire hull below the waterline.

Any breaks in a line indicated low spots or divots on the hull's surface where the paste had not contacted the wood. Skilled hands with the adz then began the careful removal of traces of the orange marks, cutting down the high spots.

Over the next few days, this procedure was repeated several times until the finished hull was smooth and fair as any modern glass hull. Magic? I thought so.

Now, here in our leaving Satawal in solemn ceremony, Lewis was given a similar amulet. He was clearly moved by the gift. It was also clearly an honor to be given such magic, but it was clear that Lewis didn't have a clue what it really was.

Mike McCoy was the last man over the side for the ride to the beach, in the skiff, as we prepared to depart. His crew was in high spirits (and drunk on Red Label) as they rode away. McCoy was mouthing the words "magic." Was this a RF or what? I meant to talk to Lewis about it but never got the chance.

We were soon headed for West Fayu and an elemental encounter that changed the character and purpose of our voyage to one of survival.

West Fayu

There is only one pass through the reef at West Fayu. It lies at the opposite end of the lagoon from the anchorage and landfall. The distance between the pass and the island was only three or four miles. Perched on top of the reef to the northwest, high and proud, appearing to be steaming straight into the lagoon, loomed the wreck of the freight carrier *Ishiru Maru*. This is exactly what she was doing the night she went aground, ahead full, with a load of Datsun pick-ups in her hold.

The captain and crew were reported to be drunk on *sake*. They were saved, but by the time the salvage crew got there to save the trucks, the locals had already stripped them of everything they could lash to their canoes. The *cargo cult*[18] was alive and well! On our cynical side we PCVs used to joke, "Pray for a typhoon or a shipwreck" (surely the ship will come, loaded with exotic stuff that people desire).

It is truly an ill wind that blows no good.

There was nothing to salvage of the cargo of the *Ishiru Maru*. The tires, batteries, wheels, and anything else that could be unbolted from the load of Datsuns had already been dispersed all over the Western Pacific by canoe. I was looking forward to taking the dingy over there in the morning to check out the remains, but this was not to be. We had to check out early.

Tri Amora was anchored quietly in the lagoon. Three canoes were drawn up on the beach and cocooned in palm fronds (wrapped so that the sun would not dry and shrink the planks and open the caulking in the hull): the turtle-gathering parties from nearby islands that we had hoped to meet. Everyone else on-board *Tri Amora* was crashed, but I was excited and restless, so I took the dingy into the beach where a party of men were feasting on some of last night's catch. More turtles were scattered around the edge of the camp, helpless on their backs, trussed and alive, ready to be loaded onto the canoes for the trip home.

Turtle is delicious just off the fire and, of course, I was asked to "come and eat." Turtles are easy to catch, where they can be found. You just wait for them to come ashore during the night to lay their eggs and you grab 'em. Trouble is there aren't many places where they can be found. West Fayu is one of the few.

Any and all turtles that come up onto inhabited islands are immediately captured and eaten, and since they are usually taken before they lay their eggs (which the Micronesians also love to eat), there are no turtles born to come back to those places like Satawal, Puluwat and Pulusuk.

The kids on Pulusuk were always organizing overnight "turtle hunts" to the far side of the island, but I think they were like "snipe hunts." Obviously, something else was going on outside the watchful gaze of the adults, who knew what was going on anyway. I never saw any turtles come back to Pulusuk.

The Shark

―⚯―

*"…When that shark bites, With its teeth babe,
Scarlet billows, start to spread…"*
-Bobby Darin-

―⚯―

It's not uncommon to encounter sharks in the lagoons of Truk. I saw them almost every time I went into the water, sometimes more of them and closer than I was comfortable with. Sometimes they appeared to be aggressive. I had seen Gray Reef sharks in a posture with back arched, jaw dropped, pectoral fins straight out and circling me at great speed, but at a distance. This is when I got out of the water: fast! I always thought the behavior was territorial. It meant "go away." It worked. Diving companions had been "bumped" by sharks with no more consequences than having to change their shorts. I can't imagine it feels especially good to be "bumped" by an animal covered with skin, the scales of which actually become teeth in a different part of its body.

From the turtle party on the beach, I learned there are sharks in the area, probably lots of them, and they are dangerous at this time when the turtles are coming in. "Don't go in the water near sundown," I was warned.

Bill did not get this message.

It was about sundown when I left my friends at the beach and began to pull the dingy back to *Tri Amora*. I didn't know that Bill was in the water scouting for incoming turtles.

Bill didn't see it coming.

It clearly took him by surprise. He was coming up toward the surface from a free dive in twelve feet of water. He was rotating his body as he came up. I find myself doing the same thing when I am free diving. I bet we all do. As he turned, it was suddenly in his face, probably a small Grey Reef shark, its jaw dropped open, back arched, and pectoral fins straight out, as Bill described it. It was coming fast.

It's hard for smart animals like us to figure out dumb animals like these, but if I'm right about the posture of the shark, it may have been in a territorial mode. It didn't want to eat Bill. It just wanted him to go away. I think Bill had put up his arm to ward off the shark. It slashed his hand, with open jaws, as it came over his shoulder raking his back with its teeth. It didn't bite.

James Stewart, the chief diving officer at Scripps Institute of Oceanography, who certified me as a diver, had a similar encounter with a small shark somewhere in the Pacific, and had a wonderful scar on his arm that he was quite proud of. It was his badge of honor.

I was standing up in the dingy alongside *Tri Amora,* ready to tie off, when I heard Bill screaming. The sound was primal, a sound I had never heard a human make. I pulled the oars furiously to cover the one hundred, or so, yards between us. He was lucky. Minutes earlier

THE SHARK

I would have been on the beach, a much greater distance away. Minutes later I would have been below deck. We did not even know he was out there and even if we had heard him, it would have taken longer to reach him.

The thing is, around here in America where I live now, we are all in danger, extreme physical danger, all the time. But we don't feel close to it. There are guardrails on every curve. The FAA inspects our planes. There are alligators in my pond, but there is a sign that says, "Danger! No Swimming." Still, thousands of us are killed on the highways of America every year. We are bored with drive-by shootings. Police killings don't happen to us. We watch buildings collapse on television. It seems so distant. It's not happening to us. Here, in the islands one is always close to danger and there are no guardrails.

You are allowed to put yourself in danger, and you know when you are in danger, so you have time to enjoy the rush. If you are a junkie for it, you will like life in the Outer Islands. I think Carlos was an adrenaline junkie.

We were all about to cough up another one of our lives on this day. Bill, maybe two.

There was blood in the water and the shark was still circling after this elemental encounter. Bill literally flew into the dingy. He was blind with rage that *this* senseless animal would do this to *him*. This expression of anger surprised me. Does this mean that I expected something else? What would I feel? To Bill this was not terror, it was war. Fight or flight. In hind-sight, and given what we were about to go through, fight would be sustaining.

What Bill said, or asked next, after the rage, was also surprising, "My face? "My face"!! (The emphasis was in his voice). "Is my face OK"? At the oars, I was looking right at him. Clearly to me his face was

OK, but I thought, "is this what I would be worried about at this moment"? His right hand was laid open, tendons severed, pouring blood. Bill had instinctively grabbed his wrist, compressing the arteries to staunch the flow. What he couldn't see was his back. It looked like a plowed field, from which red flowing rivulets trickled down into the bottom of the dingy that was now filling with blood and seawater[19].

"I DID NOT HAVE LIFE THREATENING INJURIES, nor were my hands ever in danger of having to be saved."
-E-mail from Bill Curtsinger to Diane Strong-

"My need [was] nothing more than, warding off infection and reducing pain and fever until I could get to an antiseptic medical facility."
-E-mail from Bill Curtsinger to Diane Strong-

That was Bill's assessment later. My own assessment of his situation at the time was somewhat different than his. He seemed to think he was still in the Navy and the helicopter was coming. Conditions were different here. His life *was* in great danger. In fact, all of our lives would soon be put in danger.

Bill became quiet as we closed the last yards between us and *Tri Amora*. His thoughts? Wrath against his fates and what had been wrought? His beauty and appearance at risk? I was thinking it was getting dark, there was a reef between us and open sea, and thousands of miles across open ocean to anywhere that was at all meaningful in terms of help for Bill.

THE SHARK

As I pulled the oars, I could see, silhouetted against a backdrop of flaming sky, the dorsal fin of the shark that had attacked Bill. It kept apace of us at several yards distance, slicing a spreading quicksilver gash in the reflection of a blazing red sunset. I thought the shark seemed now strangely to be an escort, and more strangely, I even found a bit of comfort in the presence.

A Shaman told me there are animals that can help us when we are helpless, or terrified, or lost. He said we should find out who this animal is.

Was there a spirit who has helped me in this way? A spirit from another realm, mostly unaware of me in this one, but who answers when I cry out. Someone who senses a disturbance in the boundary that lies between us, but outside of our ordinary senses, at a place where we are joined.

I wondered what an animal might be that knows me in this way. What is my totem? this primal spirit who I might call, who would hear me and know it is I? Who would come to my side and do battle for me against all terrors that come for me?

It is strange to think that this animal might be one who embodies that terror for me, in its realm. And I am terror for him, in my realm, as my brothers tear his fins from his body and dump him back into the sea to die. How can I help him and save him from this terror?

What would the shark know of me, here, or I of him, as we are met in the place where the sea is a wrinkled mirror and a boundary that keeps us safe from one another? Here, at this tenuous edge, where from above comes his greatest peril- me, and from below my most feared peril- him.

BEYOND TIME

I am the lord of earth and sky, ruled only by lust and desire; and he the lord of sea and slime, following only the pull of tides and faint tremors in the currents sweeping out of the depths, the deepest on earth. They come swirling among the reefs of paradise to an unfelt pull of an ancient moon, a moon perhaps itself born of this place, torn from the Earth. Is gravity its longing to return?

Over the Reef

Aboard *Tri Amora*, we took stock; we were at West Fayu, essentially an uninhabited desert island. We will soon be trapped in the lagoon by darkness. The sun was sinking fast and by the time we reached the pass we would not have visibility to make passage. We were aboard a sailing vessel. We have lost our radio and have no communication. NO ONE KNOWS WE ARE HERE. We are thousands of miles away from any antiseptic medical care. Bill's needs were urgent.

The tide was high, so that one could argue that maybe, just maybe, we could go over the top of the reef, an action that under normal circumstances would *never* be considered. True, *Tri Amora* was shallow of draft, and this is part of what made her such a great boat in this part of the world, but ironically, that is exactly how she finally met her end some years later: smashed on a reef. The *Ishiru Maru,* was an immediate reminder of disaster, wrecked on the reef right in our faces. Did we need to go tonight? We have a medical doctor to care for Bill. The morphine from the medical kit seemed to make Bill more comfortable. What about infection? We had antibiotics in the medical kit, but they are not very effective against Micronesian bugs. I had a running sores for two years from coral cuts despite anti-biotics. The amount of torn skin on Bill's back and shoulder was massive and open to infection.

BEYOND TIME

While Dr. Lewis and Ann were tending to Bill's wounds, three of us talked in the pilothouse: Carlos, Tomoichi and I. As skipper of *Tri Amora*, it was Carlos' call. He didn't want to wait twelve hours until another daylight. He wanted to attempt sailing over the reef. I thought it would be attempted suicide. Bill would get help sooner, but all our lives would be effectively put at risk. *Any* decision would be fateful, and our options were closing out. Carlos decided. Nobody took issue. It was done.

In the dying light, we headed for the reef under power.

I remember straddling the bow of the center hull of *Tri Amora*. In the closing darkness, we rode up the face of each breaking wave over the reef, and then plunged down the backside into a mine-field of coral heads. I tried to signal to the helm in hopes of us avoiding any one of which could tear us asunder. There was no turning back, and it occurred to me that we may be close to dying in this farthest of places imaginable. Not only could I see it coming, but also it was a choice. An adrenaline rush.

Carlos was at the helm and I was having trouble giving him hand signals and staying in the boat. It was like the mechanical bull ride at Gillies. I needed another hand. In truth it didn't matter which way I pointed, or that my free hand was simply waving at the heavens most of the time. Yippee eye ky ay! The coral heads were everywhere, within inches of breaking up *Tri Amora* as she plunged forward through one huge wave after another breaking on the reef. I may as well have been blind, and so also Carlos, at the helm.

Was it the hand of God that carried us over that reef at West Fayu, or did we ride over it on the back of the shark? I still felt its presence in my groin. The question haunts me still. How are God and the shark related to one another, and they to me in some divine intervention? For surely that's what it was.

OVER THE REEF

It was dark in the eastern sky when we cleared the reef. We and the shark parted company. The shark turned back to follow its prescribed course in nature. We put up sail and set a course for Lamotrek, 40 miles distant, turning south into the growing darkness. Was there help there for Bill? Was there a radio on the island? Was there anybody to hear us?

No safety net. We were on our own.

We drove fast, southerly on a reach across the constant Trades, under all the sail we could get up. I took the helm from Carlos who wasn't grinning this time. He went below. Ann had retired to the aft cabin, seasick. Lewis was somewhere below tending to Bill, and Tomoichi was asleep behind me on the pilothouse bench. It was dark in the pilothouse, just the red glow of the compass. There was not a sound except the wind and sea. I was in the Now, Beyond Time.

I found that when I was on a proper compass heading for Lamotrek, a cross of stars became visible, low on the southern horizon. I could sail by the stars. The island of our dreams, the island of our need, lay somewhere under the stars, the *Southern Cross*[20].

As the constellation, after making a low arc in the sky, disappeared again below the southern horizon sometime that night, I was still driving. The compass glowed red in the darkness, our only guide now that both the shark and the stars had abandoned their duty to lead us to safety. I wondered what was possible to know, beyond our analog portrayals of this world? I would learn more about that, after even more of our devices had abandoned us.

Lamotrek

After sailing all night, we made Lamotrek, the island closest to West Fayu. In the morning, David and Carlos went up to the radio shack. Ann was recovering from another ordeal of seasickness and I was cleaning up the boat. Bill was standing next to me on the deck, his shoulder, back and hand bandaged, arm in a sling. I turned over the dingy. We were both startled by the sight. It was full of blood mixed with sea water. Bill looked at me, but nothing was said as I washed it into the lagoon.

David and Carlos came back from the radio shack and said the Navy was sending a plane from Guam. "What are they planning to do?" I asked. "Drop a diver into the lagoon" was the answer.

I was thinking "What the hell do we need a Frogman for?" Then I thought, "This is going to be entertainment." The locals paddled out a fleet of canoes and waited to pick this guy up when he parachuted out of the plane into Lamotrek Lagoon. What an RF this was going to be.

The Lockheed turbo-prop Orion came over the island low just above the treetops, deafening at about four hundred knots. The *yat* were going nuts. It really was a beautiful sight as the Lockheed winged out over the lagoon and all the canoes were silhouetted against the afternoon sun sparkling on the water; but how was this going to help Bill?

LAMOTREK

Lewis was back on the radio asking the same question of the Navy as the Lockheed returned at altitude for the jump.

A true lifeline, the jet runway at Yap, and the Navy hospital in Guam were as far away as ever, and this wasn't going to get Bill any closer, as hard as we wished. Someone came to their senses; but when the Lockheed left without discharging its Frogman, we were nevertheless desolated. What now?

Bill had two possibilities left. Hundreds of miles to windward, back to Truk with *Tri Amora,* or an act of the Micronesian Congress to turn around the field trip ship headed for Yap (it had just left Lamotrek a day before our arrival).

Somehow the *M/V James M. Cook* got turned around. The islands of Woleai and Ulithi didn't get their mail, and Bill and David Lewis were taken straight to Yap, 500 miles distant to the east. The trip would take three or four days at most. As things turned out it would take us many more days of hardship, sailing west, to get back to Truk. I will always be thankful Bill was not with us, but rather safe on his way to care and home.

We said farewell to David and Bill. They were on their way to safety. For us it would be a different story.

The flag chief of Puluwat, Manipi, who had been visiting Lamotrek, joined us for the return (there are two chiefs, traditional and "flag" who represents Puluwat to the government in Truk). He wanted to go back to his home island, not to Truk. Manipi was a chief and Carlos had to agree. It became Carlos's responsibility to get Manipi to Puluwat. Carlos understood that to be unable or unwilling would be to lose face. Carlos understood the Micronesian concept of face. It is what sent the Puluwat navigators to Guam and him along with them.

BEYOND TIME

We set a course for Puluwat, when going to Truk would have been the rational choice.

Sometime that first night out of Lamotrek trying to power into the wind and a heavy swell, we realized that we were making no forward progress whatsoever. The engine droned on, but we weren't moving. Perhaps we had lost our prop, or the gear box had failed. Our plans to power directly to Puluwat were finished. Rather than being able to set a rhumbline course, we would have to tack under sail into the wind. Things were going to get a lot tougher than we anticipated.

We had begun our journey without consulting magic to conjure best sailing conditions and assure safe passage. Magic would have told us to wait a few days for the wind to turn around, and to have the patience to wait. Time, however, in our Western way, was more important. We had no awareness and, had we, would have ignored it anyway. Sailing conditions were not ideal. The easterly Trades still dominated, though diminished. We would have the wind on the nose, all the way. Had the winds turned, it would have been easy.

Besides not having consulted magic, and one ignores the Magic at one's peril, we started home with a couple of other significant handicaps. Though we had a compass to give us direction, it could not tell us our position. We started our voyage with a taffrail log (a towed device that measured and recorded our speed through the water), but it was lost sometime during *Tri Amora's* downwind leg from Puluwat to Satawal. We had no way to measure our speed and therefore, how much distance we had traveled with any accuracy.

We had also lost our radio antennae, it having gone overboard from the top of the mast in a strong squall. We had no communication. There would be no maydays sent out from *Tri Amora*. We were on our own.

Dead Reckoning

"FYI, the boat did not have an accurate timepiece onboard for navigation and the captain, Carlos Viti asked to borrow my Rolex so he could navigate (by sextant) his way back to Truk. He used the expertise and experience of David Lewis to get the boat out to the Caroline's but once we left to board the mail boat to Yap, he was on his own. It took months for my watch to work it's (sic) way back to me in Maine."
-E-mail from Bill Curtsinger to Diane Strong-

At dusk on the second day of the voyage, under sail in a weak easterly and heavy swell, Carlos and I tried to get some star sites with the sextant. There was only a small window of time, just after dark, to catch a star near the horizon below a high ceiling of stratus clouds, that otherwise obscured the sky, and would for many days.

Micronesian navigators are very familiar with these same stars we were looking for, which rise, arc across the sky and fall on the horizon each night.

BEYOND TIME

To the Micronesian navigators, the 360-degree horizon is like a vast compass card of thirty-two asymmetrical points at the edge, marked by the rising and setting points of certain stars: Aldebaran, Antares, Altair, Vega and others. These are called the "Voyaging Stars." Certain destination islands, or other navigation signs are known to lie under each of the rising and setting points. Star courses from any island to any other form a vast matrix which every Micronesian navigator has memorized by the time he is twelve years old. I didn't know very much about this, even though in fact I had just discovered it for myself - Lamotrek lies under the Southern Cross from West Fayu. That was an easy one.

To a Micronesian navigator, the vast panoply of sea and sky is full of signs. There are signs everywhere – If you know how to look, know how to feel, how to hear, how even to smell the fragrance of land that carries far to sea, and how to attune all your senses to notice all of these signs.

Micronesian navigators are taught from a young age to float in the sea and learn to feel the waves. It is said that a blind man can navigate a canoe by the feel of waves, in the motion of the canoe, through his testicles. I was blind to it all--awareness I didn't have.

The Micronesian system of navigation is what we call "dead reckoning." In simple form, it is the system I used when navigating racing yachts off the California coast in my relative youth. My tools were a binnacle compass, a hand bearing compass, a chart, eyesight and a few visible signs such as lights and buoys. It was how we got *Tri Amora* to Satawal and beyond (without, of course, visible signs). Those passages were easy, simply setting a direct compass course and sailing the rhumbline. Off the wind, no tacking was involved.

Micronesian navigation is a much more vast and complex cognitive system, in which they have many more tools than we: thirty-two stars, wave patterns, enumerable other signs, awareness and *Etak* (Moving Islands) - a concept that is hard for the Western mind to understand. To begin with, the Micronesian navigators imagine, when voyaging, that their canoe is stationary, and the destination island is moving toward them! In the mind of the navigator, the only fixed elements in their system are the canoe and the stars. Everything else is moving.[21]

The difficulty of our passage back to Puluwat, from Lamotrek, was immensely complicated by the physical character of the destination. Truk would be hard to miss. Its high islands can be seen from a great distance so that even if we had not accounted properly for leeway, drift, or compass deviation, we surely would pass by close enough to see it.

On the other hand, Puluwat was a different story. Puluwat was a much more difficult landfall.

Because of the low-lying nature of atolls and the curvature of the earth, they are visible on the horizon at a distance of no more than *seven* miles. Micronesian navigators extend that range of visibility by knowledge and practice of many other elements of their sophisticated dead reckoning system. These include the movements of sea birds, presence of sea life, clouds, ability to locate and recognize screening shoals, patterns of waves bent around seamounts, and other marks in the sea and sky that we don't notice. They will even compute angles and direction from the intersection of waves from different directions generated from various patterns of weather thousands of miles distant, and they have *Etak*. We did not have these tools.

Puluwat itself can be seen with the eye from a slightly greater distance due to the old WWII control tower built by the Japanese when Puluwat had an airfield. It towers above the treetops, but you would certainly have to know where to look.

BEYOND TIME

As I've said, the correct thing would have been for us to go directly to Truk, especially after the engine faltered on the first night of the homeward journey, when it made the most sense to change course. As rational as that sounds, it was *not* the right thing to do from the point of a Micronesian. Carlos understood this. Manipi, the chief, wanted to go to Puluwat. Carlos could never admit to Manipi that he was unable or unwilling to carry out this responsibility in the way intended.

Carlos was as much Micronesian as any White man will ever be, but he was a Western navigator, on a Western boat. In the West, technology compensates for our lack of knowledge and awareness of the natural world. When this technology fails, as it often does in Truk, we are in deep Kim Chi. More of our technology was to fail us.

Assurance of fair sailing weather means choosing the time to voyage. The Micronesians, understanding the windward limitations of their canoes, voyage west to Truk only in the months when the easterly Trades diminish and the winds, for a period of time in the summer months, reverse and blow from the west. Then they wait until the easterly Trades return, for the trip home.

It was the time of year when one could expect the wind to change direction. Perhaps, had our little expedition gone according to plan, and not been cut short by an angry Grey Reef shark, and had the wind turned around, we would have had a sleigh-ride home, just as we had in the opposite direction going down. But almost is not the same as certain. As it turned out, we made some wrong choices and had some really bad luck.

We were going to have to tack back and forth, and back and forth, upwind all the way back to Truk, in a boat that didn't go to weather very well. I think that this experience will always affect my choice of boats. Better to go to weather than go fast. We had immediate choices to make: do we make a few long tacks or many short tacks.

Either way, without the means to accurately know our distance traveled (speed), it was going to very difficult to keep track of where we were, and thus which direction we should go. We chose to make more, but shorter tacks, which should at least keep us closer to the rhumbline (shortest distance to our destination) at all times, but on each tack more uncertainty was introduced.

And so, I began to plot our tacks (zigs and zags) on the chart. Carlos was skeptical of its benefit, saying, "We don't know how much distance we've covered on each tack."

"True," I admitted. "We'll have to guess." Maybe a better term would have been "estimate," not that it would be any more accurate for giving it a different name. Each hour, the helmsman would note our compass course sailed, and "estimate" our speed. At the end of my watch, I would tabulate all the collected entries and plot a zigzag line on the chart which represented our course.

At the same time, I tried to take comfort in the sight of Carlos lashed to the mast with his sextant pointed at the horizon. I wanted to believe that he could pull this off.

I stood with Carlos, or should I say hung on, as *Tri Amora* pitched and rolled, holding the time cube in one hand and the mast with the other, marking the time at each site taken. Inside the cube was a lovely woman's voice, with just a hint of British accent that would, after a series of pretty chimes, repeat over and over, "Greenwich mean time is ...." Sometimes, when off watch, I would turn on the time cube just to hear the pretty chimes and her soothing voice--and dream. I don't remember the Rolex watch that Bill gave to Carlos.

I was skeptical that we could even properly identify any of the stars we could see, or get proper declinations from the heaving deck, but Carlos exuded confidence.

BEYOND TIME

After going to the Site Reduction Tables and reducing the sites, our plots were invariably far from my zigzag line on the chart, or else the triangulation of sites was too large to be of use. We could be anywhere.

When Carlos finally gave up the star sights, I realized he had no facility with celestial navigation and the sextant. My own experience with celestial navigation was extremely limited. I had taken a course but never seriously practiced it and I knew nothing about the stars in the southern sky.

I think my doubts about Carlos arose even before we left Truk, when Lewis told Carlos that navigation "would be in our capable hands" and seeing the look that came over his face. I knew when I looked at him that he was not going to admit that he was unable. To have expressed any doubt in his own abilities in front of Lewis would have been a monumental loss of face that no Micronesian man could suffer. I think that Carlos was as much Micronesian as any White man can be, and in an elemental way, he knew the meaning of face in the culture.

As days passed, I kept drawing my zigzag line. Carlos paid no attention to my chart and my own doubts about it grew. Carlos began to withdraw, to spend most of the time below deck. He became incommunicative and the grin was gone. He had quit trying to use the sextant altogether.

I don't know how many days we were at sea. I never did. I didn't keep track of the days. Time meant nothing. There was only the Now. I kept only the rhythm of the watch schedule; three hours on and six off. I was especially in the Now when I was on the helm. There were only constant corrections of course and the trim of the sails to attend to. It was like a meditation.

Whale Watching

There were a couple of surprisingly recent yachting magazines on the boat, probably left by some prior charter. I read the article about an English couple adrift on a raft in the Pacific for 270 days before rescue, after a whale rammed and sank their boat (a much stouter one than ours).

If I had not read that story, I would have seen the approaching pod of Humpbacks as a pleasant distraction.

Soon they were all around us. I couldn't count them all. The sea was in turmoil. I imagined again the ordeal of that couple in that raft. I looked down to see a whale longer than *Tri Amora* break the surface between the hulls of the trimaran. It was longer than our boat. The others swarmed nearby, their massive dark shapes breaking the sea into many faceted, mirrored, broken shards of light.

At the helm, I turned around as the great head of a whale erupted from the sea, not ten yards astern of me. It rose out of the sea, "spy hopping" until a great round eye, big as a saucer, was looking me square in the face. We looked at each other for what seemed like minutes, the whale standing motionless on its tail. I said to myself and to the whale, "we are not here to harm you." Tomoichi was frantically emptying the fuel tank of the portable generator into the sea. He knew what he was doing. The head sank slowly back into the sea and it was gone. They all were gone. I looked at Tomoichi. He was trembling and white like me.

Near Death

On average, a person can survive without water for about three days.

Ann had not taken any food, or *water,* from the start of the homeward journey (how many days ago?). Death may have been just around the corner for her. Among the many things that none us of knew was how to care for someone in this condition. What I did know was that the longer it took us, the closer to death Ann was getting. Our progress was slow, and given Ann's condition, excruciatingly so. The Trades from the east had begun to die out, but the hoped-for westerly winds, in reverse, never filled in. There wasn't much wind at all. We were in the *Doldrums.*

Ann was down in the aft cabin, her condition worsening. It would have been a cozy place to snuggle in some other clime, in some other time. Now it was just cramped and stifling. Ann never left it. I had never before, and never since, known of seasickness so extreme as affected Ann. She had long stopped asking for water. She couldn't keep it down. She would barely rouse to a touch. Her skin was red, burning hot, and dry to the touch. I wanted to heal her, but I didn't know how. The cure was so simple: to be at rest on a placid lagoon, the cure that seemed distant as the Moon as *Tri Amora* heaved, and then dropped into the trough of the next giant swell, making little progress. I felt like her life was in our hands. I didn't know what to do.

Ann was dying and I felt my zigzag line on the chart was taking us to nowhere.

A Dream

How does one dream when lost, almost without hope? I flew above the sea, around the world to a friend who was hurt. How did I know? How did it happen? Was that a dream? Was I traveling in another dimension?

Just in front of me She stood, reaching out, as toward the source of everything we want. Her bare arms shone in the stage lights, Her bare shoulders, Her bare back. I wanted to touch them all; touch Her. I did.

She never spoke or turned, but as our outstretched hands met, our fingers entwined, Her rings and my scars, we danced, together, in the music. All of Her, and the music, came through me in waves. I felt Her in every part of me, and the music. Then confusion; is this a dream? Should I feel ashamed? Everything in me, I wanted to pour out through my fingertips, and into Her. I wanted to throw myself in Her and die in Her. Obliterate myself, be one with all the waters of the world, and the Universe. I wanted to heal her.

I wanted to be a *shaman*; take her demons into myself. She was sobbing and I tried to console her. How long can we remember a dream? I still remember every detail of that encounter, as though it was real. We met a few years later and she told of events in her life on that day, just as I had seen them in the dream. She didn't know I was there. I couldn't heal her. She died of a heroin overdose, in Hollywood.

A DREAM

Still traveling, above the clouds, I could see forever. I don't know how long I was gone when, looking down through a break in the clouds, I could see *Tri Amora* among their shadows. She appeared stationary and the world around her swirling and moving. It was just as the Micronesian navigators envision. I felt a sensation of knowing exactly where we were. The ocean was plaqueless, spotless and without any track. There were no islands to be seen.

Suddenly I awoke and felt the hard deck of *Tri Amora* against my skin. I saw Manipi at the rail staring down into the depths.

We were near the position on my chart where Puluwat should be. I didn't know whether to shorten the tacks so as not to miss the island or lengthen them for a wider search. Finally, I gave up tracking altogether, tacking aimlessly. How was one going to find a nearly unseeable speck on a vast ocean?

We had no course to Ann's salvation. We were LOST.

Salvation

Manipi, as I would learn only later, was an initiated navigator of the Weriyeng school. He would spend much time at the windward rail searching the sea in rapt concentration. One morning I joined him at the rail. He was staring straight down into the sea. We didn't speak and I too stared into the sea. All I could see were streaks of light streaming out of the depths. I thought it was just sunlight. Perhaps it was something else that only Manipi knew.[22] I saw nothing. If Manipi saw a sign, he didn't say. He never once offered any corrections to our course. This would be in keeping with the etiquette of the Puluwat navigator. This was not his vessel to steer. He would leave that to us, and though we thought ourselves to be lost (because we had no idea where we were), upon reflection later, I am almost certain that Manipi knew where he was at all times. In keeping with the navigator's code, he would step in only if we made an egregious error.

On a following morning, as I awoke for my watch, Manipi called me to the rail. He pointed to the horizon, abeam. It took me awhile to finally see it, appearing like the stubble of a whisker, barely a tic above the horizon. It was the tower at the old Japanese airfield on Puluwat, poking above the still hidden trees. He said we must tack – the only command he ever gave us as a navigator. We would lay it on a single tack. Salvation for Ann was, at last, close at hand.

SALVATION

It was early afternoon as we approached the pass to Puluwat lagoon and bore off to a close reach. We started to fly. "Speed Boats" (outboard powered skiffs of local construction) came out of the lagoon to meet us. They couldn't catch us. We were flying. Manipi was at the rail shouting, waving. We were all yelling. Ann popped up through the hatch in the aft cabin for the first time in (how many?) days. I thought of Bill and hoped that he had found a faster way to the help he needed.

As we entered the pass in the lee of Puluwat Island, the wind slacked. We took off some sail and ghosted into the inner lagoon. Women in paddle canoes came out to meet us. The *yat,* in bright red or bright blue *thu,* were in the trees trilling and calling. Soon there were barebreasted girls climbing all over the boat. They brought us turtle, and fresh sweet *taro,* and cool drinking coconuts.

We slowly sailed into the inner lagoon of Puluwat, where the great hunkering roofs of canoe houses of all the Puluwat clans stood along the shore, a canopy of radiant green palms behind them. I saw Ikuliman, most ancient of all ancient mariners, tattooed head to foot, earlobes stretched to his shoulders (how fierce he must have looked as a young man), squatting on the shaded beach smoking a cigarette. All the Puluwat voyaging canoes were drawn up on the beach well-tended, cocooned with palm fronds. Their reflections were mirrored on the lagoon, smooth as glass.

Ann was already in the galley cheerfully baking brownies! You could smell 'em (where she found the box of mix, I had no idea). It was amazing how quickly she recovered from near death. Being at rest in a quiet lagoon made all the difference. We had made salvation, our spirits lifted.

Later in the evening, I was able to speak with Cheryl on the island radio. She told me she had met Bill at the airport in Truk a few days earlier, on his flight back to Hawaii. He said he was well after having been treated at the Naval Hospital in Guam.

BEYOND TIME

I pictured her in the radio shack, a closet really, just inside the swinging screen door entrance to the Maramar Hotel. I imagined the sky outside aflame over Udot, The air breathless calm. Solander is behind the bar. LoveMe, in a flower print island dress is serving the tables on the screened porch and humiliating anyone who makes the slightest proposal to her. The evening's entertainment had begun. 'Shakin' Aetkin is railing against the hippy, Commie Peace Corps volunteers, his table already stacked with empties. Two or three Peace Corps volunteers are at the next table comparing the morning's stool sample.

The Korean guy is winning the hearts and minds of the Trukese, teaching them the basic points of the martial arts, and buying the beer. He was extremely personable, and I liked him. There was always a crowd of Trukese young men around him. They loved the warrior shit he was teaching them. He was perfect for his job with the fishing company that wanted Truk's fish. He had gone to school at Cal Poly SLO (a couple of years before my time there) and was a Karate Champion, but never made it Kung Fu in Hollywood. He arrived in Truk with the Korean long-liner now at the dock down by the Fishing Co-op. They didn't bring any fish, nor did they take any—this time. There were two new Toyotas lashed to the cover of the forward cargo hold. A few days later, I would see them again, in lean-tos next to each of the Senators' homes in Mw an Village. The Koreans would get all of Truk's fish they wanted.

Cheryl said Bill Curtsinger was appreciative of the care and help we had given him at West Fayu. What he told her was the first news she had heard of us in the many weeks. It must have been a shock. On the call, I did not tell her of our ordeal after leaving Bill on Lamotrek. That story would wait for a gathering of the clan. We were still three days from Truk, which lay to the east, wind still on the nose, There was no other way home for Ann. As soon as *Tri Amora* began to feel the swells of open ocean, she was down for the count. We were still out there, Beyond Time.

The Island

We left Puluwat with little fanfare and made for Truk. After two or three more days at sea, we arrived outside the reef at night, guided in by the lights of the radio towers on Moen. We hove-to for the night, not wanting to try the pass in the dark. One more night of Hell for Ann.

It was another crystal day next morning, just as when we left so many days ago, as we sailed through the Northwest Pass and broke out the bottle of California Cabernet.

Why do we voyage? For the leaving, for the arriving? Or simply, as once explained by the great Puluwat navigator, Ikuliman, "to get away from a nagging wife to a distant place of complacent women."

Legacy of a Peace Corp Volunteer

Shortly after my return from the Western Islands, Cheryl and I prepared to leave for home in America. The leaving was bittersweet.

———⋈———

I took away much more than I left behind. This will be true of any good Time Traveler. Do no harm. Leave not a trace.

———⋈———

-My thoughts homeward bound aboard an Air Micronesia 727 somewhere over the Pacific, longitude unknown, late summer in the Northern Hemisphere, 1973.

Epilogue

Carlos Viti died tragically on the island of Oahu in 1996. Hit and run. He was on a bicycle. He didn't see it coming. No time for an adrenalin rush.

Carlos was described by Dan Baker, an author and filmmaker who knew him well, this way: "Carlos Viti was an important American photographer in the post-war Western Pacific. His work chronicled the Micronesian people and their lives during the nation-building times of the Trust Territory government. His work effortlessly transcended the ordinary and depicted an unknown people in a faraway place with respect, understanding, and a special piercing insight. His Portraits of the Palau Chiefs series for the Smithsonian have been compared to the work of Edward Curtis, who produced magnificent photographs of Native Americans from 1907-1930.[23] Carlos wanted the Micronesians to speak for themselves, without analysis or filters. Carlos strove not only to give them a face to the world, but a voice as well."

Hipour died a natural death tended by his family as he lay upon the cool, spotless tile floor of his home on Puluwat. His son became a navigator, following in his father's footsteps.

Ikuliman died in 1983. At the age of thirteen, his grandson, Manny Ikea, accompanied his grandfather, Ikuliman, on the "renaissance" voyage from Puluwat to Guam in 1972. After Manny was initiated in the Pwo navigator initiation ceremony on Puluwat in 1997 by Hipour, he changed his last name from Ikea to Sikau in honor of his grandfather, Ikuliman Sikau. Manny died in 2013[24].

David Lewis, in the later years of his life, became blind but continued to sail with the help of friends. He died in 2002, in Queensland. One obituary said of David Lewis that "He was a typical Polynesian sailor, getting into trouble through haste and neglect, then, with near superhuman courage and seamanship, fighting his way out of it. Beautiful and intelligent women were drawn to him. He leaves three wives, four adult children and many, many friends." I think David and Carlos were essentially cut of the same cloth.

Tri Amora met a tragic end, was never salvaged, after having been driven onto a fateful reef somewhere unknown. In her place are now a dozen state of the art "dive boats" discharging swarms of tourist divers onto the wrecks of the *Ghost Fleet* and hosting festive "shark hunt adventures" (with bang sticks for those inclined to kill dumb fish for the simple fun of it).

Yosua, Cheryl's lover, was killed in a bar fight in Honolulu.

Killer Miller was lost in Laos, in a war that no government, to this day, will admit happened, fighting for the wrong (?) side.

Micronesian navigation and canoe building are still practiced today in the Caroline Islands of Micronesia. Life in the outer islands of Truk and Yap is little changed in the fifty years since this story began. The renaissance of traditional voyaging that began in 1972, as I accounted here, continues and has gained worldwide attention. The *Pwo* ceremony was performed by Hipour, in 1997. Before that, Urupiy performed a *Pwo*[25] ceremony on Lamotrek in 1990, "for the first time before Western eyes" which was filmed by

EPILOGUE

Eric Metzgar and co-produced with Urupiy's son, Ali Haleyalur. The title of the completed film is *Spirits of the Voyage,* from Triton Films.

I am encouraged by reports that traditional voyaging canoes are still being built in the Carolines today.[26] This is due, in part, to the fact that inter-island ship traffic between the eastern islands of Yap and the western islands of Truk is still nonexistent, and motorboat travel is unreliable and costly. It is due also to the pride of Micronesians in their heritage. That heritage is also a tribute to the enduring ingenuity of humankind and deserves to be more widely known. Such a tribute was afforded to the Micronesian navigators by the renowned filmmaker and deep-sea explorer James Cameron who stated, "The best inspiration I got for *Avatar 2 and 3* was dealing with the master navigator culture in Micronesia."[27]

The island of Tamatam continues to shrink. Someday it will no longer support a human presence. The voyaging canoes will carry this fragile community to a new and distant place,

―∞―

"Riding Mother Nature's Silver Seed to a new home in the Sun"
- Neil Young

―∞―

The *Cargo Cult* appears to be alive and well in Truk.

―∞―

WENO (Moen), Chuuk (Truk) During the ongoing regular session of the 7th Chuuk State Legislature, Senate Joint Resolution 7-12 was

BEYOND TIME

introduced to "respectfully requesting his Excellency Fidel Castro, President of the Republic of Cuba, to give special attention to the plight of the people of the State of Chuuk Federated States of Micronesia as a result of the United States' neglect toward the needs for sustainable economic development of the islands."

-Resolution urging Cuba to give 'special attention' to Chuuk-
By George Hauk for
Marianas Variety News And Views
Friday June 20, 2003

———⚯———

Life in the Outer Islands is unchanged. The people are proud, resourceful and independent, Beyond Time and certainly politics.

I returned to America and entered a time warp (and culture shock). The first movie I saw was "American Graffiti." America had changed after the turmoil of the '60s. The first car I bought was a screaming yellow '56 Chevy with a hot rod motor. It was time to get on with life; me and Corporal Niguchi (a Japanese soldier who never surrendered, found on Guam in 1973 stealing chickens).

The Sea Turtle will return to the shifting fringes of Pikelot and West Fayu for generations and be fruitful. They may return to Tamatam, when it becomes uninhabitable, after all the people have left, out there **Beyond Time.**

———⚯———

EPILOGUE

Once In A Lifetime

You may find yourself living in a shotgun shack
And you may find yourself in another part of the world
And you may find yourself behind the wheel of a large automobile
You may find yourself in a beautiful house, with a beautiful wife

You may ask yourself, "Well, how did I get here?"

Letting the days go by, let the water hold me down
Letting the days go by, water flowing underground

Into the blue

-Talking Heads-

References: Micronesian Ethnography and Voyaging

Gladwin, Thomas. *East is a Big Bird: Navigation and Logic on Puluwat Atoll*. Harvard University Press, 1970.

Krause, Stefan M. *Preserving the Enduring Knowledge of Traditional Navigation and Canoe Building in Yap, FSM*, Voyaging and Seascapes, 2016.

Lewis, David. *We, the Navigators,"* The University Press of Hawaii, 1972.

Lewis, David. *The Voyaging Stars, Secrets of the Pacific Island Navigators*, W.W. Norton & Company, 1978.

McCoy, Michael. *A Renaissance in Carolinian-Marianas Voyaging*, The Journal of the Polynesian Society, Vol 82 1973, p. 355-365.

Metzgar, Eric. *Carolinian Voyaging in the New Millennium*, Micronesian Journal of the Humanities and Social Sciences 5(1/2) (2006): 293-305.

Ridgell, Reily. *The Persistence of Central Carolinian Navigation*, A Journal of Micronesian Studies Volume 2, Number 2 / Dry Season 1994, p. 181-206.

Photographs

Scenes depicted here in 1972 and 1973 would appear much the same today. The photos are by the author unless noted otherwise.

CARLOS

MARAMAR HOTEL

CHERYL

TRI AMORA:
under full sail on Truk lagoon

HIPOUR:
with permission from Harvard University Press, from
"East is a Big Bird", by Thomas Gladwin.

IKULIMAN:
Photo by Carlos Viti, with permission of his estate

SANTY:
Pulusuk navigator and master canoe builder

TOMOICHI:
Faithful crew and companion on the voyage of the *Tri Amora*

DAVID LEWIS (L) and BILL CURTSINGER

ANN VALENTINE

PULUSUK: CANOE BUILDING

LAMOTREK: CANOE HOUSES *(WUUT)*

MAGIC: Divining the time for voyaging.
Photo by Carlos Viti, with permission of his estate

TAMATAM: DANCER
with a *maramar* in her hair

PULUWAT: GREETING PARTY
upon the return of *tri amora*
with their chief, Manipi

PULUWAT: return from Pik,
Ikuliman and turtle

PULUSUK: dispensary construction    THE AUTHOR

WEST FAYU: *Ishiru Maru* on the reef    PULUSUK: dividing the catch

PULUSUK: trolling for tuna

PULUSUK: my welcome party. Inocenti (left), my *pwim* (friend) for life

PULUSUK: Volunteer, Katie, dressed out island style

PULUSUK: Father Fucky's church

PULUSUK: *Truk Islander* anchored on the drop-off

PULUSUK: Michael and my construction crew

PULUSUK:
Juli, Michael's daughter

PULUSUK:
Okapi, my spear fishing partner

PULUWAT: returning voyaging canoe

YALEILEI
SATAWAL: ANCIENT MARINER
The stylized dolphin tattoos on his legs are the
mark of navigators (*palu*).
Photograph by Nicholas Devore, with permission of his estate.

**LAMOTREK: GOONY BIRD CHICK PERCHED
IN FRONT OF CANOE HOUSE.**
Birds, when seen at sea near dusk, invariably fly straight
toward land. This one, when mature, will lead the
navigators right back to their front door!

LAMOTREK: TRI AMORA AT REST
after the shark attack on Bill, waiting for evacuation

PULUSUK SCENE

WEST FAYU

VOYAGING CANOE

THE NEIWE HOUSE: Our Peace Corps home for two years.

LEAVING TRUK FOR THE LAST TIME

PEACE CORPS POSTER
CIRCA 1970

End Notes

1. The Caroline Islands (or the Carolines) are a widely scattered archipelago of tiny islands in the western Pacific Ocean, to the north of New Guinea, on the Equator. The Carolines are scattered across a distance of approximately 3540 kilometers (2200 miles). They are part of the larger Remote Oceania, the part of Oceania settled within the last 3,500 years, comprising Island Melanesia south and east of the Solomon Islands archipelago, plus the open Pacific: Vanuatu, New Caledonia, Fiji, Palau, Micronesia, and Polynesia. That is the dictionary definition. When asked for his definition of Remote Oceania, Eric Metzgar told me "Out of sight." Indeed.

2. Truk Atoll is an atoll in the central Pacific and part of the Caroline Islands. Truk is about 700 miles northeast of New Guinea located mid-ocean at 7 degrees north latitude. Today it is called *Chuuk,* having reverted to its ancestral name upon independence in 1978. I use the old name, "Truk" in my story simply because that's how I remember it. Similarly, I have used other island names as I remember them. Puluwat is now *Polowat*. Moen island in Truk Lagoon is now *Weno*.

3. Truk was the Japanese Pearl Harbor of WWII when, in February 1944, waves of American bombers attacked and sunk more than 150 Japanese warships to the bottom of the lagoon. They remain there today, the *Ghost Fleet*, and attract wreck divers from around the world. Otherwise, the deep clear waters of Truk Lagoon are the realm only of *ghosts,* those of the Japanese soldiers and seamen who perished there.

4. *Taro* is a root that is pounded into a paste, smothered in sweet coconut milk, and served on a pandanis leaf. To tourists in Hawaii, it is known as *poi* and is best eaten with the fingers.

5. Hipour was one of the most famous and accomplished of Carolinian master navigators.

6. Hipour had indeed been to Saipan, when three years prior, he navigated David Lewis's modern steel ketch *Isbjorn* to Saipan, without instruments, using only the directions passed to him from ancestors in oral tradition.

BEYOND TIME

7   For an understanding of this migration, across the vast Pacific, from out of Southeast Asia to Easter Island, I recommend reading *The Voyaging Stars*, by David Lewis.

8   *Thu:* The traditional loincloth worn by men in the Outer Islands.

9   After sea trials, newly constructed canoes are painted with a smooth black finish, to ease passage through the water, and are trimmed in brilliant red or yellow.

10 The Proa is a double-ended outrigger sailing craft which always carries the outrigger to windward as a counter-balance weight to the overturning force of the wind on the sail. When changing direction in relation to the wind, the ends of the vessel must be reversed in order to keep the the outrigger on the windward side. This process is called shunting as opposed to tacking, where the sail is simply shifted from one side to the other.

11 Carolinian voyaging canoes are very sophisticated in design, utilizing principles of aero and hydro dynamics not understood in the West until later times. A significant aspect, and advantage, of the Carolinian design is that since the same side is always to windward, the hull can be made axially asymmetrical. That is to say one side of the hull can have a different shape (curvature) than the other. Why would we want to do this? It's the same principle that lifts airplanes, and when applied to a boat hull it can lift the canoe to windward allowing it to sail closer to the wind. The leeward side of the Carolinian canoe (facing away from the outrigger) is made straighter (flatter) than the windward side (toward the outrigger) which is made to have more curvature. As the canoe travels through the water, the water flows past the hull faster on the straighter side creating a high-pressure zone. On the curved side, where the drag of the curvature slows the flow of water, a low-pressure zone is created. The hull is literally "lifted" by the high pressure into the low-pressure zone, which is to windward. This is just as a wing, with more curvature on the top, lifts the airplane. This principle was not understood in the West until the early twentieth century, making flight possible. That the Micronesians had put it into practice, centuries before, is a testament to their intelligence and ingenuity.

12 Neiwe is a small village on the south end of Moen island and the site of a WWII Japanese sea plane base, a portion of the ruins of which our house was built upon.

13 This practice was recorded by Carlos Viti in remarkable photographs he took on the voyage from Puluwat to Guam with Ikuliman. Eric Metzgar calls the practice not really magic, but a form of divination called *pwe* where young coconut leaves are randomly tied in knots to foretell the future.

14 The Mariana Trench, at 35,000 feet in depth, is the deepest place on Earth. It was created by the collision of two great tectonic plates. The Pacific Plate, wandering west, collides here with, and dives below, the Asian Plate. Unimaginable forces

END NOTES

have shaped this gash in the Earth's Crust and pushed up the Himalayas, the tallest mountains on Earth.

15   *Amora:* The Micronesian name for the sail of the voyaging canoe.

16   Jacques Cousteau came to Truk, found and explored the wreck of the Japanese submarine I-69 and made a movie about it that is very dramatic and haunting. The I-69 sunk during the raids in 1944 when someone left the hatch open as she dived to escape the bombing--80 sailors perished. She had not been touched since the war. While I was in the Westerns, Buddhist priests ceremonially came to collect the remains and return them home. Their ghosts however still remain, as any Trukese knows.

17   The Carolinian navigators call it the *big wave*. It is one of three used for navigational purposes. This wave is generated by a great circulation system around a high-pressure area known to sailors racing the Trans Pac to Hawaii as the "Pacific High" far to the north. The winds flowing away from the center of this high toward relative low pressure surrounding it are bent into a counterclockwise gyre generated by the Coriolis force of the earth's rotation. The result is an easterly wind on the southern edge of the gyre while at the northern edge the flow in the north Pacific is the opposite, or westerly. These winds are called the trade winds, or simply, the *Trades*. The easterly Trades blow unobstructed across the entire Pacific, generating very large waves from the east in their path. When the high pressure diminishes in the summer months, the winds flow in reverse, from west to east, in Micronesia. We were counting on this reversal for our return to Truk.

18   The term *cargo cult* originated in indigenous practices in the Melanesia subregion of the southwestern Pacific. After WWII, the cults would practice ritualized building of elaborate structures constructed with tree limbs and vines looking like rustic topiary frames mimicking runways, aircraft hangers, and control towers in order to attract airplanes that were hoped to come and disgorge mountains of Western goods upon them. Every human being wants an easier and better life, but something is lost, and something is gained in living every day. This story is about what may be lost.

19   Bill Curtsinger's personal account of the attack can be found in his wonderful book of photographs, *Elemental Nature, p. 160.*

20   I kept the position of the Southern Cross which lay in the sky just outside the portside shrouds as a reference to steer by until it set below the horizon. The Carolinian name for this position of the Southern Cross is *Machemelito* (Southern Cross 45° setting-it lies on its side) . It is part of their *Paafu* (sidereal star compass) and is applied to the *Wofanu* ("looking at the island") compass for setting a course from various islands in the Caroline archipelago. Lambda Scorpii (also known as Shaula), in the constellation Scorpio, is the star used by navigators for Lamotrek when one is on West Fayu. In the traditional Carolinian navigation system, it is called *Tubula Metariue* (in Lamotrekese) meaning "Setting Shaula."

These notes are from personal correspondence with Eric Metzgar who, among many accomplishments, was initiated in the *Weriyeng* school of traditional navigation on Yap in 2015. The difference in the Carolinian's course to *Tubula Metariue* and mine to *Machemelito,* which lie nearby in the sky, could possibly be due to their accounting for more leeway made by their canoes than for our modern sailboat.

21 *Etak* are reference islands that lie to the side of their course. They are unseeable but known to lie under certain stars. As David Lewis describes it, as the destination island 'moves' toward the canoe, the reference islands are also 'moving.' In other words, the canoe is conceived as stationary beneath the star points, whose position is also regarded as fixed. The sea flows past and the island astern recedes while the destination comes nearer, and the reference island moves 'back' beneath the navigating stars until it comes abeam, and then moves on abaft the beam. This is the essence of the concept—that one *Etak* along the course corresponds to the apparent 'movement' backwards by one star point of the reference island. Naturally the Carolinians are perfectly well aware that the islands do not literally move.

We can call this a figure of literary style, a canoe pictured pushing through the sea with everything moving past it except the stars poised overhead. Gladwin says, "For the Puluwat navigator, it is not a matter of style. It is a convenient way to organize the information he has available in order to make his navigational judgments readily and without confusion. This picture he uses of the world around him is real and complete. All the islands which he knows are in it, and all the stars, especially the navigation stars and the places of their rising and setting. Because the latter are fixed, in his picture the islands move past the star positions, under them and backward relative to the canoe as it sails along. The navigator cannot see the islands, but he has learned where they are and how to keep their locations and relations in his mind." I think of it as better than a hand-bearing compass because it references un-seeable things outside one's course. The Micronesian system of navigation is entirely rational, in the mind of the navigator.

For anyone interested in this subject of Micronesian navigation, especially if you are a sailor, I recommend reading *East is a Big Bird,* by Thomas Gladwin, and *We, the Navigators,* by David Lewis.

22 In fact, as I was to learn later, *sea lightning* does mean something. It is a sign the Micronesian navigators know. Biologists believe it is phosphorescence emitted by minute organisms, when disturbed by faint refracted wave impulses from shoals or islands.

23 Dan Baker, *Eulogy for Carl 'Carlos' Viti.*

24 Eric Metzgar, *Carolinian Voyaging in the New Millennium,* Micronesian Journal of the Humanities and Social Sciences 5(1/2) (2006): 293-305.

END NOTES

25 *Pwo* is the ritual initiation ceremony for Navigators.

26 Eric Metzgar, *Carolinian Voyaging in the New Millennium*, Micronesian Journal of the Humanities and Social Sciences 5(1/2) (2006): 293-305, and Stefan M. Krause, *Preserving the Enduring Knowledge of Traditional Navigation and Canoe Building in Yap*, FSM Seascapes.

27 Rebecca Keegan, *James Cameron: Avatar' sequels to draw on 'master navigators*, Hero Complex, Los Angeles Times, March 30, 2012. Cameron learned the power of the Micronesian culture while on the island of Ulithi during his ultimately successful attempts to dive, in a submersible, to the bottom of the Mariana Trench at a depth of 35,000 feet. He was successful after two failed attempts only after magic was employed in a voyaging ritual performed for him by the chief of Ulithi.

www.ingramcontent.com/pod-product-compliance
Ingram Content Group UK Ltd.
Pitfield, Milton Keynes, MK11 3LW, UK
UKHW041503030325
4829UKWH00035B/518